QUANTUM EVOLUTION

QUANTUM EVOLUTION

JOHNJOE McFADDEN

W. W. NORTON & COMPANY
New York   London

For information about permission to reproduce selections from this book, write to
Permissions, W. W. Norton & Company, Inc., 500 Fifth Avenue, New York NY 10110

The text of this book is composed in Postscript Linotype Minion
with the display set in Castellar
Manufacturing by The Maple-Vail Book Manufacturing Group

Library of Congress Cataloging-in-Publication Data

McFadden, Johnjoe.
Quantum evolution / Johnjoe McFadden.— 1st American ed.
p. cm.
Includes bibliographical references (p.) and index.
ISBN 0-393-05041-6
1. Evolution (Biology). 2. Quantum theory. I. Title.

QH366.2 M396 2001
576.8—dc21·   00-53320

W. W. Norton & Company, Inc., 500 Fifth Avenue, New York, N.Y. 10110
www.wwnorton.com

W. W. Norton & Company Ltd., 10 Coptic Street, London WC1A 1PU

1 2 3 4 5 6 7 8 9 0

To the memory of
Roly, Tom and Alice

CONTENTS

ILLUSTRATIONS

ACKNOWLEDGEMENTS

This book would never have been written were it not for the support and encouragement of many friends and colleagues. Special thanks go to those who read and commented upon drafts of the manuscript, particularly Martin Adams, Jim Al-Khalili, Tanya Baron, Michael Conrad, Greg Knowles, Chris Nunn and Malcolm von Schantz (but the blame is all mine). To Greg and Jim I owe particular thanks for the numerous inspirational conversations. Georgina Laycock did a heroic job in converting my rough-hewn manuscript into real writing. I owe particular thanks to Philip Gwyn Jones and Toby Mundy for their enthusiasm, support and encouragement. I owe most to my wife Penny for being my most thorough reader, my fiercest critic and greatest friend. Lastly, I must thank Ollie for constantly reminding me of what is really important in life.

QUANTUM EVOLUTION

1

What Is Life?

Starlight glistens on a spaceship's silvery hull as it cruises, unseen and unmanned, amongst the planets of a distant solar system. Guided by the encoded instructions of an alien civilization it glides past dark, rocky planetary outposts and bloated gas giants until it reaches its goal, and swings into the orbit of an inner planet. A probe is released. Retrorockets fire that adjust the probe's trajectory, easing it slightly from the mother-ship's geostationary orbit and turning its heat-resistant nose towards the ground. The grip of the planet's gravity drags the probe inwards, through ever-decreasing orbits. Faster and faster it spins until, plunging through clouds, it finally emerges under a leaden sky. A parachute is released to halt the headlong dive, and the craft slowly descends to land on a rock-strewn landscape.

Minutes later, a metallic lid is drawn back, exposing a camera lens, and pictures are beamed back to the mother-ship. The camera pans across the rocky scene. The same rubble-strewn landscape is everywhere – rocks of all shapes and sizes lie sunken into fine grey sand. The air is still. Nothing moves. The camera scans the monotonous surface stretching in all directions towards the horizon – grey rocks, some precariously balanced atop others, others lie shattered, blasted by the forces of alien weather. The camera pans again, and then one rock, in shape and colour much like any other, spreads its wings and soars into the sky. The mother-ship sends a signal backward through the vastness of space, towards the distant home of the spaceship's makers: LIFE!

The planet is, of course, Earth and the rock a bird, perhaps a rock pigeon, lost in barren desert. The story illustrates the wonder we should feel at the most remarkable phenomenon in the known universe – life. Our telescopes and space-probes return images of the universe's many marvels – the twisted braids of Saturn's rings, Neptune's moon Mirander's scarred and shattered surface, the birth of stars within the Crab Nebulae. Extraordinary as these are, they pale before the astonishing nature of life itself. And yet, all life forms are essentially rocks – made of the same materials, obeying the same laws, as the rocks, stone and sand that surround us. We are rocks that run and swim, climb and leap; that hear, touch and see; rocks that can look out into the vastness and grasp for an understanding of ourselves and the universe that made us.

In this book we will explore the nature of life and ask what animates living organisms. What is, in the words of Dylan Thomas, 'The force that through the green fuse drives the flower'? To understand the nature of this *force*, we must explore life at its most fundamental level, examining the two key events in Earth's history that made the act of writing these lines possible. The first took place nearly four billion years ago, when life emerged. The second took much longer. Living creatures had been swimming in Earth's oceans for three and a half billion years before the mammals gave rise to a family of bipeds, the primates, and from their ranks emerged a thinking ape, man. Since that time, several million years ago, the mind of man has unravelled many mysteries concerning the universe's workings. We watch the sun setting every evening and are confident of its rise the next day, because we know its rising and setting are caused by Earth spinning on its axis. We can look up into the night sky and know that each star is a sun like our own. Scientists can calculate the energy released from the fusion of hydrogen nuclei inside our sun, or use powerful telescopes to witness the birth of galaxies that existed billions of years ago. Remarkably however, the two key events that made our own existence possible – the emergence first of life and then of consciousness – still remain mysterious. Although we know now a great deal concerning both the workings of living cells and (though far less) the human brain, the spontaneous appearance of both phenomena remains a puzzle. This book's aim is to explore this puzzle and examine the startling proposition that we already hold a missing piece of the puzzle. We will discover how, with this piece in place, enigmatic phenomena can be explained and light shed on life's central mysteries.

To approach the answer, we must first understand the meaning of the question. What is life? What *is* the force that through the green fuse drives the flower? Living today inside the concrete and glass walls of urban environments, it is easy to ignore life's astonishing nature. Our perceptions are formed within homes shared with domesticated animals and potted plants and only slightly modified during weekend excursions across forest-denuded hillsides or through fields of monoculture crops. The forces of the natural world are often perceived as problematic: mould creeping over damp patches of bathroom walls, weeds encroaching on flowerbeds or ants invading kitchens. But it is in our encounters with these weeds and vermin that we glimpse nature's true character. The moulds, plants and insects invading our homes and gardens are heirs of the creatures that first colonized the oceans and proceeded to relentlessly invade every habitable niche on this planet. If we are to unravel life's secrets, it is their nature we need to understand.

One starting point is to examine how our ancestors, unsullied by the preconceptions of our civilization, viewed their world. Man first walked on the planet several million years ago. For almost all subsequent history, man's chief preoccupation was the gathering, snaring and hunting of nature's bounty. Our ancestors' day-to-day survival was contingent upon the ebb and flow of life through their landscape: the migration of herbivores, the ripening of fruit and the spawning of fish. To survive, man needed to exploit these resources, and he learned to lay traps to catch animals, grind tools to butcher them, fashion clothes from their skins and kindle fires to cook them. But the same reasoning that endowed *Homo sapiens* with his unique skills to exploit nature, condemned him to remain for ever discontented with mere exploitation. Man sought to understand his world. Our ancestors held nature's procreative power in awe, worshipping gods and goddesses whom they represented as sexually exaggerated figures – such as often heavily pregnant 'Mother Earth' figures (FIGURE 1.1) or priapic males. Life's vitality was celebrated in the vigorous images of bison and reindeer that leap across the cave walls at Lascaux or Altamira (FIGURE 1.2). These two aspects of nature – its energy and its capacity to reproduce – clearly impressed our ancestors, and still remain mysteries of life today.

Much of subsequent history is a reflection of the changing pattern of man's interaction with the rest of the natural world. After several million years as a hunter and gatherer, man turned his skills towards manipulating nature. About ten thousand years ago – apparently independently in

FIG 1.1   The Venus of Willendorf [c. 30,000–
25,000 BC; limestone sculpture; height: 11 cm;
Lower Austria]. © Naturhistorisches Museum Wien,
Photo: Alice Schumacher.

several parts of the world – people discovered how to cultivate grain and domesticate wild animals. Man thereby freed himself from a perpetual march in search of a moving food supply, and established settlements. A surplus of plentiful crops allowed the rise of an aristocracy, who hoarded and guarded this resource. This enabled many to escape from the drudgery of tilling the land altogether. Warriors and servants could be paid from the royal coffers and thus persuaded to protect the lands of their kings, to build walls or erect palaces and temples. The level of social organization required for these tasks was previously unknown amongst the hunter-

FIG 1.2   Palaeolithic cave painting of bison from the Lascaux cave,
France [c. 17,000–15,000 BC]. © N. Anjoulet, CNP – Ministère de
la Culture, France.

gatherer communities and a remarkable invention was devised to keep
track of their transactions. Symbols and signs were scratched onto clay
tablets, representing bales of wheat, jugs of beer, or heads of cattle,
either paid to, or received from, the king's subjects. From these modest
beginnings, writing developed. Information and ideas encoded on baked
clay tablets could be faithfully transmitted across space and through time.
Fortunately, those ancient scribes turned from recording the jugs of beer
paid to their workmen to more interesting information: the beliefs, hopes
and dreams of their people. The stories they tell are our first detailed
records of man's thoughts concerning life.

The earliest creation myths record the belief that life represented the
fundamental creative power in the universe. The universe's origin was
itself often held to be some form of birth. In the Orphic creation myth,
black-winged Night laid a silver egg in the womb of darkness; Eros was

hatched from the egg and set the universe in motion. Similarly, the *Rig-Veda*'s Hindu creation myth describes the birth of the first being from a golden egg, all other deities springing from his limbs. The authors of these myths were mostly farmers. and much mythology revolved around the seasonal cycles. They sowed their fields with seed and marvelled at its power to sprout and grow into luxuriant crops. Their myths reveal that they generally traced this power to a divine source. The ancient Sumerian sky-god Enlil is described as:

> The lord (Enlil) who brings forth what is useful
> The lord whose decisions are unalterable
> Enlil, who brings forth the seed from the earth[1]

Life is clearly considered to be apart from the rest of creation, its vitality a channelling of divine power. The cycle of growth, death and rebirth was, within agrarian societies, almost universally attributed to the death and rebirth of a fertility god or goddess. Thus, Osiris, the Egyptian god of vegetation, was said to have been slain and dismembered by his brother. His wife, Isis, gathered together his body's scattered fragments and with magical ceremonies restored him to life. The Egyptian reapers chanted a dirge for the death of Osiris and prayed to Isis for his return. Similarly, the descent of the Babylonian goddess Ishtar into the netherworld echoed the desolation of the dry season; her subsequent rescue and emergence restored the growing season's fertility. The cycles of human fertility were seen as under divine control. The coincidental synchronicity between the moon's waxing and waning and the female menstrual cycle was attributed to the influence of a lunar deity, such as the Roman goddess Juno, to whom barren women would pray. The (less obvious) role of male reproductive organs in procreation was also recognized. Thus Aphrodite was born from Uranus' testicles which had been flung into the sea by his son Cronos, who had castrated him with a saw-toothed sickle.

Each tale records a belief that life contained a divine or magical principle, absent from the inanimate world. To create life, this vital principle needed to be added, often from a living source, such as blood. Thus, in the Babylonian Poem of Creation, it is related how man was fashioned from clay mixed with the blood of a god:

> 'Let him be made of clay animated by blood'.[2]

Although, today, it is easy to dismiss these myths, they are in reality man's earliest attempts to find answers to the questions still plaguing us – they are the first theories of everything. Today we know where the sun goes at night and why spring follows winter. But much of our knowledge is received wisdom and this wisdom of ages was hard-won. How many of us would be able to *prove* that the Earth revolves around the sun, when any fool can see the sun rise in the morning, travel across the sky and descend below the horizon at night?

The dawn of the rational approach to understanding our world is usually attributed to the intellectual revolution of the sixth and fifth centuries BC which gave rise to the ancient Greek civilization. One of the earliest philosophers was Thales (born about 600BC). Although his writings have been lost, several of his sayings have survived, including, 'the lodestone has life, or soul, as it is able to move iron'. This short phrase implies a complex set of beliefs. Firstly, that the ability to initiate movement is a key attribute of life. This is a concept we will return to as, in modern molecular interpretation, it forms a cornerstone of this book. Secondly, that this ability to make movement betrays the presence of a 'soul'. Like the mythmakers before him, Thales considered that the phenomenon of life pointed to the presence of supernatural forces. Finally, the equation: ability to initiate movement = life = soul, has been taken to the extreme of attributing the property of life to a variety of inanimate objects, such as a magnet (lodestone). This reflects a widespread tradition of pantheism in the ancient world. As the third-century Roman chronicler, Diogenes Laertius put it, 'the world was animate and full of divinities'.

The ancient world's greatest biologist was undoubtedly Aristotle. Sadly, our received image of him is frozen by those chalk-white busts of venerable bearded philosophers who seem to stare into a perfect world of spheres and equilateral triangles. But Aristotle's vision was far more earth-bound than that. Like his predecessor, Heraclitus, he believed that 'knowledge enters through the door of the senses', and as a young man he spent several years living on Lesbos, studying marine life. His biological writings betray the acute observation and attention to detail which is the hallmark of all great naturalists.

'Animals also which fly and those which swim, fly by straightening and bending their wings and swim with their fins, some fish having four fins and others, mainly those which are of a more elongated form (eels for example), having two fins. The latter accomplish the rest of their movement by bending themselves in the rest of their body, as a substitute for the second pair of fins. Flatfish use their two fins and the flat part of their body, instead of the second pair.'[3]

Instead of the venerable sage, we should imagine a younger Aristotle diving into the Aegean's clear waters to retrieve starfish, crabs and anemones, to study their form or observe their behaviour.

'The sea-urchin has a better defence system than any of them: he has a good thick shell all round him fortified by a palisade of spine.'[4]

Any lover of rock-pools will recognize an ally in Aristotle's writing. But the scientist in Aristotle was not content to describe nature; he needed to explain it. Perhaps, later in the day, he would set light to driftwood to cook his catch and ponder on the ephemeral quality he roasted out of the living flesh. Like Thales, Aristotle considered that the essential quality of living creatures was that they possessed their own internal *will* and this allowed creatures to initiate independent movement.

'For nature is in the same genus as potency; for it is a principle of movement – not however in something else but the thing itself.'[5]

To Aristotle, living creatures were made distinct by their ability to *move* themselves. His concept of movement was more subtle than simple locomotion. The shoreline of Lesbos, had taught him that clams, anemones, or indeed simple seaweed moved very little (except when pulled by the waves and the tide), but were still very much alive. To Aristotle, there were six forms of movement: generation, destruction, increase, diminution, alteration and change of place. This broader conception of movement actually reflects a more general meaning to the verb, *to move*, than our modern usage, one that remains apparent when we say that we found a particular piece of music to be deeply moving, or when a motion is passed by a debating society. Our modern usage is rooted in Newtonian mechanics, and a better translation of Aristotle's concept of movement would be the term *action*, a word with a precise, useful meaning in modern physics, to which we will return. The essential point of Aristotle's argument is that all living organisms possess an internal will that allows them to initiate and perform actions such as growth, regeneration, procreation and movement. Aristotle, like Thales, ascribed this internal will – the cause of independent action – to the *eidos*, the soul or *psyche*: 'The soul creates movement'.[6]

It would be mistaken to equate Aristotle's *eidos* too closely with the Christian soul. He believed all animals and plants were endowed with a 'soul' capable of initiating movement. To Aristotle, this soul was clearly a much more functional entity than the Christian moral guardian. However, only man possessed the highest form of soul: the source of reasoning and moral judgement.

Aristotle's writings, lost and then found by the Arabs and passed from them to mediaeval Europe, were to form the basis of Western thinking throughout the Middle Ages. The Aristotelian concept of a *soul* was translated into the *vitalist* approach to biology. To the vitalists, life possessed a mysterious property, the *élan vital*, or living spirit, whose nature lay beyond the realms of science. In the words of Joyce Kilmer:

> Poems are made by fools like me
> But only God can make a tree.

The vitalist tradition survived until the twentieth century in many biological writings. I remember biology textbooks that described the mysterious living protoplasm inside cells with the same awe and mystery that mystics describe the aura. However, the concept has been in retreat since the dawn of the Age of Reason in the seventeenth and eighteenth centuries, and no serious scientist subscribes to it today. The opposing camp, the Mechanists, were inspired by the machines that were, by then, revolutionizing the world; and they believed that life, like machines, could be understood in terms of the laws of chemistry and physics. They rejected the vitalist argument that life required special laws beyond conventional science. René Descartes (1596–1650) was a founding figure who proposed that animals were mere automata, in principle no different from the clockwork figures which played music or danced at fairgrounds. Descartes was however unwilling to accept the full implications of mechanism and reserved man a special place amongst God's creations. He considered man's intellectual capabilities, his reasoning power, betrayed the presence of an immortal soul. Mechanists had to wait for another century before books such as La Mettrie's *L'Homme Machine* (Man the Machine) (1748) laid bare the full force of the mechanist manifesto. La Mettrie agreed that animals were no different from machines but argued that man differed from animals only in complexity. The way was now open for science to delve into the very substance of life.

Technical advances in analytical chemistry and microscopy naturally drove the life sciences towards reductionism – the belief that complex systems can be considered as the sum of their parts. Nineteenth- and twentieth-century scientists began a reductionist dissection of the chemistry of life. In 1853, the Lille brewing industry hired Louis Pasteur to discover why their wines soured. At that time, fermentation was considered purely a chemical reaction. Brewer's yeast was thought to be a chemical catalyst facilitating the conversion of the grape-sugars to alcohol: yeast was not recognized as a living organism (which is not so strange when you examine it in its powdery form). The brewers' hiring of the brilliant young chemist was, thus, hardly surprising.

Pasteur had made his name demonstrating that tartaric acid crystals came in two forms, left- and right-handed that were mirror images of each other. When he synthesized tartaric acid in the laboratory he grew crystals with approximately equal proportions of the left- and right-handed forms. However, when he extracted tartaric acid from living tissue, the crystals he grew were always left-handed. Pasteur found that the same was true for nearly all biochemicals extracted from living tissue: if the chemical came in a left- and right-handed form, then only one would be found in living tissue. Living systems were *chiral*. He was therefore astonished to find that growing crystals out of wine fermentations, he obtained only left-handed crystals. This convinced Pasteur that he was dealing with a biological process rather than a simple chemical reaction. He confirmed his suspicions by demonstrating that yeast was a living microbe that fed on sugar, generating both alcohol and (sour) acids as the waste products. He thereby discovered the cause of the souring in brewing, and simultaneously founded the sciences of microbiology and biochemistry.

This marriage of mechanist philosophy and reductionism led to the great triumphs of twentieth-century biology. Over many decades, the myriad of interlocking biochemical pathways forming the living cell's metabolic skeleton were laid bare. This knowledge and capability has led to major innovations in medicine and biotechnology. It may seem churlish to question the success of the mechanist/reductionist approach. Yet, despite its undoubted success in elucidating the chemical processes that underscore life, has it really enabled us to understand life itself? It is noteworthy that several centuries since the *mechanist* manifestos of Descartes or La Mettrie claimed that living organisms were mere machines, we have not succeeded, despite numerous attempts, synthesizing life in the laboratory. Joyce Kilmer's line still hold true. No one has ever made a tree, or a flower or an animal or an insect or even the lowliest bacterium.[7] The sole means of making life is procreation – sowing the seeds of pre-existing life forms. Though scientists can confidently describe physical chemical reactions taking place at the centre of our sun, at the surface of black holes, or during the first millisecond of the universe's existence they cannot achieve what the lowliest life forms on Earth manage with ease: make life.

Our failure to put the ingredients of life together and obtain anything living suggests something must be missing from our list. Perhaps we should start by examining why the mechanist/reductionist approach has failed to tackle life's fundamental questions: there is a paradox that lies at the heart of the reductionist approach to biology. As one dissects the workings of any living creature, examining the detail of smaller and smaller components (we will be attempting to do just this in the following chapters), life itself seems to vanish before our eyes. Whilst we have no difficulty in recognizing life in a whole animal, or indeed in one of its cells; when we come to looking at the cell's insides, the question seems to evaporate. Is a chromosome alive? What about a gene or DNA? Is a ribosome alive, or a protein or an enzyme? The question seems to lose its relevance when applied to these bits of life. The components of living cells, stripped of context, seem fundamentally no different to inanimate chemical systems. Life seems to emerge only at higher levels. It is, to use modern jargon, an 'emergent' phenomenon: one that cannot be entirely understood in terms of its parts. As a means to explaining life, the unrelenting reductionist approach is doomed to failure.

There are of course many phenomena, both biological and non-biological, that do not succumb to what the philosopher Daniel Dennett describes as 'greedy reductionism' – the mating behaviour of birds, ecology or politics – to name but a few. Each has its appropriate level of explanation and no one would attempt to analyse them at the level of fundamental particles. However, that life itself is such a phenomenon, whose appropriate level of explanation lies at cells or above, is not generally appreciated. But if we cannot hope to understand life by dissecting it, what alternative approach can we use? We should start by looking again at what we are studying. What is life? We still have not answered the question that troubled our ancestors thousands of years ago. The modern answers have suffered the fate of reductionism. Life is reduced to a collection of parts which, in isolation, have lost their essential livingness. Attempts by scientists and philosophers to identify the key properties of life read rather like a checklist: self-replication, sensitivity, evolution, heredity, metabolism, etc. etc. If you can tick more than three boxes then it's probably alive; but in isolation none is either necessary or sufficient to define life. Consider self-replication. This is generally considered to be a key attribute of living organisms. But not all organisms that appear to be alive are capable of self-replication. No mule has ever produced offspring. Many of the hybrid varieties of garden plants are sterile. Even when we examine life at a cellular level there are many cell types (for instance, nerve cells) which are certainly alive but are unable to replicate. The same kind of arguments can be used with any of the properties said to define life. They all somehow miss the key feature.

But is it so hard to identify what is alive? When I was a child, a popular children's game was 'Animal, Vegetable or Mineral'.[8] It is a simple guessing game in which the first player has secretly written down something – a dog, house, brick, carrot – indeed, anything at all. The object of the game is for the player's opponents to guess what is written down. They can ask only simple questions and the reply is always yes or no – but with one exception. The first question asked is: is it animal, vegetable or mineral? In my experience of playing this game, I do not remember any player having any problem deciding whether it was animal vegetable or mineral. We would all agree that a lion was an animal, a turnip vegetable and a brick mineral. It seems that even as children, we have little difficulty recognizing the qualities that identify and characterize living things. But how?

To make the game both harder and more enlightening, we should put ourselves in the position of the alien spacecraft; imagining that the things we have to identify are completely unfamiliar. How then would we decide what is alive or dead? How would we recognize life on other planets? The Exobiology (search for alien life) Programme of the American Space Agency at NASA has the following definition: 'Life is a self-sustained chemical system capable of undergoing Darwinian evolution.' This strikes me as an impractical life definition, particularly for an exobiology programme. How long would any spaceprobe have to wait to detect Darwinian evolution on a planet? It is hard enough to detect Darwinian evolution on Earth. I am also sure that the NASA definition was not the one we used as children; yet we were still able to identify life. So, what did we use?

Our alien spacecraft spotted the rock pigeon's ability to fly and this prompted the LIFE signal. Flying is a particularly impressive example of mobility, the property that Aristotle recognized over two millennia ago as the essential characteristic of life. However, as he argued, the mobility characteristic of life is far more subtle than mere movement. Sand-grains blown by the wind are not alive. Water flowing along the course of a stream is not alive, nor are the stones tumbling down the stream-bed. But a salmon leaping up a waterfall is alive, and instantly recognizable as of a quite different nature than either water or stones. It would not matter if an alien salmon were coloured green and shaped like a carrot; if it leapt upstream we would recognize it as living.

But what is it about the motion of a salmon or a bird that makes it so distinctively *animate*?[9] It is that a fish swimming or a bird flying is initiating its own movement against the prevailing exterior forces. Water flows towards the sea under the influence of gravity. The water's currents (mostly frictional forces) tumble a rock along a stream-bed. But the salmon's majestic leap out of the spray of a waterfall seems to defy both gravity and current to climb upstream towards its spawning ground. This seems the crux of the matter. Inanimate objects such as water or rocks are moved by the forces surrounding them; but living organisms have a internal vitality and vigour allowing them to defy these forces of nature and perform autonomous or *directed actions*.

This capability to initiate actions is both more general and more fundamental to life than mere movement. Exploring this further, we return to our alien spacecraft and imagine it has landed in a forest devoid of animal life. How would it recognize immobile plants as living organisms (we will for convenience ignore the possibility of moving plants such as the Venus flytrap)? From our argument above, we would look for a plant's ability to initiate action or movement against prevailing exterior forces. There are many ways a plant does this. The most obvious is its ability to grow. But many things grow. A mountain may grow (if you wait long enough), or a fire may grow. However, a mountain is pushed up by plate tectonics; a fire increases if the temperature of surrounding flammable material exceeds the temperature needed to ignite the material. In both these cases, the growth is in response to exterior forces. Neither possesses the ability to initiate autonomous actions. In contrast, the acorn initiates the process, culminating in the generation of a mature oak tree: it is a *directed action*. If we filmed the growth of an acorn and replayed the film, speeding the action so that the tree's entire life took just a few minutes, then we would see the oak appearing to raise itself up from the forest floor – in defiance of gravity – propelling itself towards the sunlight. This ability to move against external forces is a fundamental property of life, one lost when life is lost. If we continued to run the film of our oak tree for many years, we would observe that eventually the tree would no longer sprout new growth in the spring; it would remain leafless and eventually it would lose the ability to defy gravity, falling to the forest floor.

Somehow, whilst an organism remains alive, it is able to resist external forces and perform directed actions. When a pigeon perched on a tree decides to fly, its directed action is to beat its wings, thereby creating the turbulence that lifts it up into the air. Although inanimate objects may similarly perform actions, they lack the ability to direct them. Consider a stick of dynamite. In a sense, it can perform an action by exploding and may similarly get lifted into the air. Is the dynamite any different from the bird? Yes it is. If we determined the chemical composition of the dynamite and then added the exterior forces acting upon it, we could predict the dynamite's subsequent behaviour and the effect on its environment. We could predict when it would explode. The dynamite cannot direct its action. Its behaviour is entirely deterministic.

Determinism is one of the bedrocks of classical science. It is the principle that the future (or present) state of any system (say, the stick of dynamite) is determined solely by its past. If you know the precise configuration of any system, by adding in the laws of physics and chemistry, you can calculate its future behaviour. The principle is at the heart of Newtonian mechanics, allowing astronomers to calculate the movement of planets from their known positions and trajectories and so forecast the precise times of solar and lunar eclipses far into the future (or back into the past). It is of course entirely impractical to determine the precise positions of all particles for anything other than the simplest systems but, in principle, determinism should reign – we should be able to predict when the dynamite would explode from knowledge of existing conditions. There is nothing that the dynamite can *do*, no *action* it can take, that would affect when it is likely to explode. It does not possess the ability to *direct* its own actions.

However, if we similarly determined the pigeon's precise chemical composition and added the prevailing temperature, wind conditions, etc., could we predict that it would fly up into the air? Perhaps. But then, suppose it spied a bag of seed on the ground. It would then be more likely to descend towards the food. But perhaps there is a cat nearby. The pigeon might decide to wait in the tree until the cat has crept away. Could we predict all these possible behaviours by analysing the chemistry of the pigeon alone or even that of the pigeon and its surrounding environment? The only differences which have led to the pigeon's altered behaviour are the pattern of light photons that fell upon its retina (carrying the images of food, cat, etc.). If we include these photons in the equations of motion that describe the pigeon and its environment, would the equations predict such widely different outcomes?

I hope to convince you that the answer to this question is no. We cannot account for life with classical science alone. In particular, we cannot account how living creatures are able to direct their actions according to their own internal agenda. For higher animals, such as ourselves, we call this ability our *will*. The ability to will actions is a profoundly puzzling aspect to living organisms that appears to contradict scientific determinism. There is no role for will in determinism; we do not have choices. Every action that we perform should be determined, not by any decision we make but by the precise molecular configuration of our bodies at the time preceding our action.

So can living creatures will actions? In subsequent chapters, we will explore how all actions, at a molecular level, involve the motion of fundamental particles. Different actions will involve entirely different sets of movements of these particles. For a bird to decide to soar into the air, it must change the direction of motion of billions of particles within its body. This capability to direct motion in response to an internal will appears to escape classical determinism, and is why biological systems are so unpredictable. Its influence may even be carried over into our interactions with our surroundings. The stick of dynamite would become just as unpredictable as the pigeon, if a man was standing close by, armed with a length of lighted touch-paper. Our directed actions cause the movement of particles both within our bodies and in our surroundings.

I should emphasize at the outset that I will not be invoking any mysterious forces to account for our will, only the known laws of physics and chemistry. I am not suggesting any return to *vitalism*. Over the coming chapters we will explore how all biological phenomena – mobility, metabolism, respiration, photosynthesis, replication and evolution – involves the motion of fundamental particles. We will examine how these dynamics are governed, not by classical physics, but by the non-deterministic laws of quantum mechanics. At its most fundamental level, life is a quantum phenomenon. We will go on to explore the implications of this realization for our understanding of life's origin, its nature, evolution and consciousness. I hope, by the end of this book, you will have a new and exciting insight into what it means to be alive.

2

The Limits of Life

The makers of this alien spacecraft would hardly be content with one bulletin on a rock pigeon's flying capabilities. After its first report the spacecraft would explore further – *to seek out new life* – in the words of *Star Trek*. Its next task would be to discover what the phenomenon of life on earth actually is. What does it need? Where does it thrive? What are its limits?

Our spacecraft would soon discover that all life on Earth is *carbon-based*, that carbon is the key ingredient of our biomolecules. We might also describe life as *water-based*, since water is the substrate for our cells and tissue fluids, taking an active role in most of life's activities. Life's other main chemical ingredients are hydrogen, oxygen and nitrogen and small quantities of minerals such as calcium, magnesium, iron and sulfur.

These are readily available in our biosphere. Water is in the sea, in rivers, streams and lakes and, of course, rains frequently down upon us. Carbon is found in both inorganic molecules like carbon dioxide (CO_2)[1], methane (CH_4) or calcium carbonate $(CaCO_3)$ and as organic[2] forms such as the sugars, fats or proteins derived from the bodies of other living organisms. Hydrogen and oxygen are available in inorganic forms such as water (H_2O) or as a component of organic compounds. Similarly, nitrogen is available in inorganic nitrogen gas (N_2), ammonia (NH_3), nitrates and nitrites, and in organic compounds. Animals are unable to utilize the inorganic forms of most of these, obtaining the elements they need from organic sources – the bodies of dead plants and animals.

Life would not have progressed far on our planet if all organisms were as feeble in their synthetic capabilities as animals. Fortunately plants and microbes are much more versatile. Billions of years ago, photosynthetic bacteria[3] developed the trick of extracting carbon from the carbon dioxide in the air and stringing together the carbon atoms to make simple sugars. This is not easy; in fact, photosynthesis is one of the trickiest chemical reactions we know of (we will be looking at it more closely in Chapter Five). The problem is that the carbon atoms in carbon dioxide *prefer* to be attached to oxygen rather than tied to each other to make complex biochemicals such as sugars or proteins. To persuade carbon atoms to form complex biochemicals, bacteria and plants need a hydrogen source (plants use water) and an energy source (sunlight). Photosynthetic organisms extract carbon dioxide from the atmosphere and add hydrogen and sunlight energy to make simple sugars. The sugars are then strung together, pulled apart and reassembled to make the cell's complex biomolecules – proteins, fats, carbohydrates and DNA.

Plants did not invent photosynthesis but stole the idea from bacteria – quite literally. Chloroplasts, the organelles performing photosynthesis inside leaves are descendants of a bacteria called *cyanobacteria*. Cyanobacteria are far more ancient than plants, and performed photosynthesis on Earth at least a billion years before the arrival of plants. The ancestors of modern plants were probably symbiotic partnerships between primitive fungus-like organisms and the photosynthetic cyanobacteria, perhaps resembling today's lichens. This partnership slowly became permanent, and all today's trees, ferns, flowers and grasses are the descendants of this marriage of convenience.

Cyanobacteria are not the only bacteria to perform photosynthesis and probably not even the first. Like plants, cyanobacteria perform *oxygenic* photosynthesis – they release oxygen as a product of their photosynthesis. The oxygen comes from their hydrogen source: water. Other bacteria can utilize alternative sources of hydrogen – such as hydrogen sulfide (H_2S), ammonia or organic compounds – to fix carbon. These bacteria perform an *anoxygenic* photosynthesis which does not generate oxygen. This form of photosynthesis almost certainly preceded its oxygenic cousin.

Many bacteria and all animals are unable to fix atmospheric carbon dioxide, extracting it instead from alternative inorganic and organic chemical sources. Bacteria are the most versatile chemical feeders, able to extract carbon from a wide range of chemicals, which include organic compounds, carbon monoxide, calcium carbonate, methane, methanol, ether and formic acid. One group of bacteria using methane as both carbon and energy source is common in animals' intestines, marshes and oxygen-deficient mud. But their most bizarre habitats were discovered on the sea-bed. In the summer of 1997, Chuck Fisher of Pennsylvania State University and Phil Santos from the Harbour Branch Oceanographic Institute were in a mini-submarine, *Johnson Sea Link*, exploring the sea-bed seven hundred metres below the Gulf of Mexico. They were examining the huge bubbles of methane hydrate forming when natural gas (methane) seeps up from the ocean floor, mixing with water and other hydrocarbons to form a dirty yellow *methane ice*. Scientists had suggested that methane-eating microbes might also feed on the hydrates, but what Fisher and Santos did not expect to find was a multitude of pink worms using oar-like paddles to crawl over, or burrow into, the ice. They were a new species of *polychaete* worms. It is unlikely that they eat methane directly; instead the worms probably graze on methane-eating bacteria colonizing the ice. It has even been suggested that the worms might build burrows to cultivate farms of these bacteria.

The next ingredient for life, nitrogen, should not be a problem since eighty per cent of the air we breathe is nitrogen gas. However, we cannot assimilate nitrogen gas from air – too unreactive – we breathe it in and right back out again. Instead we obtain our nitrogen from organic chemicals in food such as, for example, the protein in meat. Plants are able to assimilate inorganic forms of nitrogen such as nitrate (a compound of nitrogen and oxygen), but this does not solve the problem since the only non-biological source of nitrate is lightning strikes which generate temperatures high enough to *burn* atmospheric nitrogen and yield nitrate.

With only very limited supplies of *fixed* nitrogen available from lightning, life might have become severely nitrogen-limited billions of years ago. Fortunately, bacteria (including cyanobacteria) discovered how to *fix* nitrogen in the air to make the soluble compounds ammonia and nitrate. Nearly all biological nitrogen is derived from these nitrogen-fixing bacteria in soil and water. *Leguminous* plants (such as peas) form symbiotic partnerships with nitrogen-fixing bacteria, allowing them to grow in nitrogen-depleted soil.

The last ingredients of life – the minerals like calcium, sodium, magnesium, phosphorus and iron – are fairly readily available, usually as salts dissolved in water. Most organisms can readily assimilate inorganic sources of these elements, such as the sodium chloride (NaCl): the salt we sprinkle on our food.

Living organisms are extremely versatile in their ability to utilize a wide range of both organic and inorganic chemicals for the elements that make up their biomolecules. Animals need much of their biomass supplied as ready-made organic molecules. Bacteria have minimal requirements: some subsist on little more than a diet of air and rock.

ICE-COLD LIFE

The average temperature in London is about 13° Centigrade, rarely going above thirty degrees or dropping much below zero. Most *higher* plants and animals are happiest within a similar range of temperatures, so it is hardly surprising that life is particularly abundant in these latitudes. Humans do live in far more extreme environments. In Timbuktu, the Saharan temperature can rise to 50°C, whilst the inhabitants of Dawson in the Yukon valley endure nights where temperatures drop to −30°C. However, even mad dogs and Englishmen would succumb to heatstroke under a Saharan midday sun and frostbite would soon freeze anyone foolish enough to brave the winter nights of Alaska. Man survives these extremes of temperature by building shelters to provide warmth or shade, thus creating a more equable microenvironment protecting him from the heat and cold outside. The range of temperature that humans can endure (without resort to ingenuity) is actually quite narrow, lying somewhere between 5°C and 30°C.

Many animals survive more extreme environments. Often considered a barren wasteland, during its summer months the Antactic is teeming with life. Millions of seabirds and sea mammals nest on its coasts and fringe of drifting pack ice. Even the snow harbours life. Warmed by the summer sun, the interior of the pack ice becomes laced with channels of slushy brine filled with photosynthetic bacteria and algae. Antarctic mites burrow through the snow to graze upon on the microscopic bloom. The summer melt releases billions of these microbes into the ocean, to be harvested by the filter-feeding krill and channelled into the food chain supporting the seals, penguins and whales of Antarctica.

Within the interior, conditions are far harsher. The coldest temperature ever recorded was a chilly one hundred and twenty-nine degrees below zero at the Russia Vostok station in July 1983. Yet Antarctica is far from sterile. It harbours more than a thousand plant species, mostly mosses, fern and lichen. The topmost peaks of mountain ranges that rise above the ice are often colonized by lichen. Indeed, brown yellow and grey spots of lichen are ubiquitous on exposed rocks throughout the world. In Donegal, Ireland, where I was born, lichen has been scraped off rocks for centuries and used to colour wool for the cloth known as Donegal tweed. The same coloured lichen spots cling to the paving stones of disused paths and cover the crumbling ruins of ancient buildings. Lichen is actually two organisms: a fungus and an algae (or sometimes a bacterium) living in symbiosis. The photosynthetic algae provide nutrients that feed the fungus. What the fungus contributes is less clear, but it probably provides support and the ability to extract essential minerals from the rock. The success of this pairing allows lichen to colonize extreme environments barred to fungi or algae alone. However, even lichen cannot perform photosynthesis below zero. Although they survive the freezing temperatures of the Antarctic winter, they must await the warming sun to heat their rock substrate to a balmy 0–10°C before they can grow and reproduce. A similar *freeze and burst* strategy is followed by most of Antarctica's flora which await rare warm spells to initiate a frantic flurry of growth and reproduction – generating millions of frost-tolerant spores or seeds – and closing down again when wintry conditions return.

The dry valleys of Antarctica are probably the most inhospitable regions on Earth. Bone-dry hurricane force winds race unimpeded across the Antarctic plateau, bringing temperatures dropping to −52°C. The winds quickly evaporate any traces of moisture from stray snow drifts. Bodies of long-dead seals and penguins lie perfectly preserved, desiccated, in conditions where even microbial decomposition is halted. A group of American scientists led by Diana Freckman of Colorado State University occupy a research station near the permanently frozen Lake Hoare in the

McMurdo Dry Valley. Studying the ecology of the area, they have discovered a curiously simple ecosystem within the soil. Though the land is
frozen half a mile deep, it is covered by a thin dry soil eroded from the
rocks by the scouring wind. This soil harbours frost-tolerant bacteria and
algae which are grazed upon by one or two species of nematode worm,
themselves the prey of a third species of worm. The worms are mostly
present in the soil as desiccated husks. Only when a rare trickle of snow
meltwater moistens the soil do the microbes and nematodes spring into
activity, hurriedly grazing, eating and reproducing before the freeze
entombs them again. Nobody knows how long the worms can endure
this Rip Van Winkle lifestyle, years certainly, but perhaps decades or even
centuries.

When the sun sets on the Antarctic summer, the temperature plum-
mets and everything freezes. Birds flee north and most seals seek warmer
waters. An exception is the Wedell seal which remains a lonely outpost
of mammalian life on the frozen pack ice (apart from the occasional
naturalist). This hardy survivor winters over in the Antarctic by using its
teeth to drill holes in the ice to the relative warmth of the ocean waters
below (at temperatures a few degrees above freezing – far warmer than
the air above) and its still plentiful food supply.

On the Antarctic landmass, covered by three miles of ice, there is no
escape from the winter cold. The lichen, algae, nematodes and mites
freeze within the snow, ice and soil to await the return of the sun. No
living thing moves.

Except, that is, for the Emperor penguin. When all other animals flee
north, the Emperor penguins head south to the freezing continental interior
where they congregate in nesting sites on the central Antarctic plateau. The
female lays her single egg and heads north herself, to the ocean, leaving
the male penguin to perform perhaps nature's most exemplary display of
paternal duty. He gathers the egg in a pouch and, huddled together with as
many as twenty-five thousand other penguins, he braves the coldest place
on Earth. For three months the male penguin endures bitterly cold temperatures, searing winds and hunger before his mate finally returns with food
for the newly hatched chick (but none for the stalwart male who must make
his own way to the sea to find his next meal).

So do the activities of the Emperor penguin represent the lower
temperature limit for active life on Earth? Not really, for the Emperors

do not *live* at a low temperature. What the colony achieves is essentially equivalent to what the town of Dawson manages to do for its inhabitants – it maintains an equable microclimate. Each penguin shuffles continually around the colony, burning his fat store, generating heat which remains trapped within the huddled mass of feathers in the nesting site. The birds are thus able to maintain their internal body temperature close to the avian optimum of 42°C, well above the freezing temperatures outside the colony. Impressive though the penguins' adaptation to the Antarctic winter is, their cells do not function at temperatures any lower than our own.

To find the lower limit for active life we must plunge into the waters below the ice of Antarctica. The world's oceans occupy two thirds of the earth's surface and approximately ninety per cent of the surface waters are colder than 5 degrees. Yet the oceans teem with life and are our most productive ecosystem. Most of the fish and invertebrates that live in the sea have body or cell temperatures that remain close to 5°C for most of their lifespan. Although the pace of life does tend to slow down at these temperatures (with concomitant longevity for many marine animals – sea turtles may live for more than two hundred years), life does thrive. Even temperatures a few degrees below the *normal* freezing point of water (salty water freezes below 0°C) are tolerated by Antarctic fish which incorporate a kind of antifreeze protein in their cells to prevent their tissues from freezing.

The coldest waters in the world are probably Antarctica's brackish pools. Don Juan Lake is saturated with about forty-five per cent calcium chloride and does not freeze unless the water temperature falls below −48°C. There is no photosynthetic activity in the lake, but live bacteria have been recovered from its waters. Whether these are true colonists or have merely drifted into the lake is unclear. They are certainly active when warmed to zero degrees. Liquid water is also found within the Antarctic Dry Valley lakes. Though the surfaces of these lakes are permanently frozen, geothermal activity can warm the deeper waters to temperatures as high as 25°C. A team of scientists drilled into these polar oases and extracted water samples that were found to contain a rich microbial flora with many unique bacterial species. The bacteria thrive just below the ice layer, where the temperatures may be as low as −2°C, and in brackish waters may drop to −12°C.

Freezing kills most living organisms. There are two ways ice damages living cells. Firstly, the mechanical shearing brought about by the formation of sharp ice crystals in cells, slices through the membranes, making the cells leaky, so that they die once thawed. Another problem arises because freezing expels dissolved salts that accumulate between the ice crystals, reaching concentrations toxic for living cells. There are, however, many organisms that can endure freezing. Even animals, particularly insects, frogs and lizards, can be frozen and thawed. Some species of frogs and turtles actually encourage the formation of ice crystals within their tissues. They make ice nucleation proteins which promote rapid freezing with the formation of smaller, less damaging ice crystals. Many microbes survive freezing due to the presence of *cryoprotectants* in their cells. Dormant life forms (seeds and spores) may survive for long periods, frozen, and have even been shown to endure temperatures close to *absolute* zero (−273°C, the temperature corresponding to a complete absence of heat – covered further in Chapter Six). The spores and seeds prevent ice formation by excluding free water from their cells. Many also produce large amounts of simple sugars that harden to form a glassy casing to protect the delicate enzymes and membranes inside. Even watery animal cells can also survive freezing. Human egg and sperm cells and even human embryos are routinely frozen during fertility treatments. The key to their survival seems to be freezing under carefully controlled conditions which minimize the damage to cells by promoting the formation of only very tiny ice crystals.

So although the frozen state is generally damaging to living cells, it does not necessarily destroy life. It does, however, prevent all living activity. Though ice and snow may shelter frozen seeds and spores and even frozen animals, nothing stirs in solid ice. Everything once alive is either dead or dormant. Active life is clearly incompatible with water in its solid state, ice. So the lower temperature limit for activity seems to be that experienced by marine and fresh water life in Antarctica, which may remain active down to about −12°C.

HOW HOT IS TOO HOT?

To explore the upper extremes of temperature, our spacecraft must leave the Antarctic to travel to the baking deserts. Life is surprisingly abundant in many desert regions. Burrowing mammals survive by engineering air-conditioning systems to keep their tunnels and hence their bodies relatively cool during the day and restrict their hunting to the cool (often very cold) night hours. Their prey – small snakes, reptiles and arthropods – may tolerate temperatures up to 50°C. Temperatures as high as 60°C have been recorded in foraging ants as they race across the hot desert sands of the Sahara. But these are transient temperatures and cannot be tolerated for long by any animal. Plants including cacti and a number of desert grasses can tolerate quite high temperatures, but do not grow above about 45°C. Mosses and lichens may survive and grow at temperatures up to about 50°C. No plant or animal is known to be able to thrive above this temperature. This appears to be the upper limit for multicellular life on our planet.

It has long been known that bacteria are capable of growth at higher temperatures. *Thermophilic* (heat-loving) bacteria, growing at temperatures as high as 65–70°C, have been isolated from a number of hot habitats. Lichen and other microbes penetrate the surface of desert rocks. Even the sand is inhabited by microbes. The surface of sand drifts is often crusty from the presence of a tangled mesh of photosynthetic microbes and lichen that live in its top millimetre. These microbial mats can be quite productive, with roughly the same density of chlorophyll as a plant leaf, but are limited by the availability of moisture and must await the arrival of dew, fog or a rare shower before they are able to set their photosynthesis machinery into action. Thermophilic microbes can also be found in more mundane environments such as compost heaps, slag heaps (that reach temperatures as high as 60–70°C), and domestic hot-water systems.

It was generally thought that temperatures higher than about 75°C were incompatible with life. This view changed dramatically when, in the late 1960s, Thomas Brock, a microbiologist from the University of Wisconsin, was walking in Yellowstone National Park. The park, famous for its hot volcanic springs, lies within a volcanic crater where rain-water seeping through the surface rocks meets the hot magma below. The superheated water and steam erupt as geysers and springs, feeding hot volcanic pools. These pools are deadly to most plants and animals. The bones of buffalo or elk and even the occasional tourist are occasionally washed up on their shoreline. Yet, in 1964, Brock noticed that the surfaces of many hot springs were covered with a pale-pink gelatinous scum, not dissimilar from the bacterial scum clinging to the inside of bathroom taps. He and his wife Louise returned the following year and isolated algae and bacteria from the hot scum. One of the bacteria, *Thermus aquaticus*, isolated from a hot volcanic spring called Mushroom Pool, was found to thrive at temperatures as high as 80°C. These *hyperthermophilic* bacteria are now of considerable interest to industry as a source of heat-stable enzymes.

The most extraordinary hot-water habitat was discovered in 1977 when the geologist John Corliss of Oregon State University and John Edmond of the Massachusetts Institute of Technology boarded the sub-marine *Alvin*. The two scientists and a pilot climbed inside a two-metre diameter titanium sphere, built to withstand the massive pressures at the depths of the ocean floor. The vessel was dropped into the Pacific, two hundred and eighty kilometres north of the Galapagos Islands to search for hot springs associated with the mid-oceanic ridges, where the continents were being pushed apart by molten magma welling up from cracks in the earth's crust. The craft (descending at a leisurely rate of 30 metres a minute) took about ninety minutes to reach the ocean floor, two and a half kilometres below the surface.

The crew stared through Plexiglas portholes to see a bleak terrain of black basaltic rock cut by faults and fissures. For thirty minutes they surveyed this monotonous sterile landscape seeing nothing unusual until a pair of large purple sea anemones drifted in front of their searchlights. The crew chased their prey over the crest of a ridge and were astonished to find themselves in the midst of a fabulous oasis of life. Sea anemones and snake-like pink fish with bulging eyes moved through shimmering warm waters, whilst crabs and miniature lobsters crawled amongst fields of giant clams and reefs of mussels. For the remaining five hours the crew took photographs and measurements and hastily collected as many of the animals as they could catch in *Alvin*'s specimen basket, before ascending to the surface.

Alvin made fifteen dives to the underwater oasis in 1977 and collected a mass of data, photographs and specimens. Since then several other expeditions have descended to discover more about the geology and biology of these unique habitats. As the team suspected, the hydrothermal vents form when seawater seeps into cracks a mile or two deep. The water is heated by hot magma to temperatures above 400°C (high pressure prevents the water from boiling), mixed with hydrogen sulfide and spewed out of the seafloor through lava-encrusted chimneys, known as black smokers. The animals inhabiting the vent live in the cooler waters that surround the hot springs. One of the most curious creatures is the giant tubeworm, which forms dense pink forests around the vents (see figure overleaf). The worms grow to several metres long but have no digestive system: no mouth or gut. Instead they depend on symbiotic bacteria that live within their tissue and utilize hydrogen sulfide as an energy source to make organic compounds such as sugars, which nourish the worms. Bacteria are present not only as symbiots but are prevalent in the surrounding cold waters and the hot walls of the black smokers. Massive temperature gradients are found within the walls of the smokers and, within the cooler zones, thermophilic bacteria flourish. The record is currently held by a bacterium named *Methanopyrus*, plucked out of a black smoker by *Alvin*, which can grow at temperatures as high as 112°C.

LIFE IN THE DARK

It is often stated that all life on Earth depends ultimately on the energy from sunlight. Plants need sunlight, animals eat plants and some animals eat other animals. But the oceanic trenches discovered by *Alvin* are thousands of metres below the ocean surface, far beneath the depths that light can penetrate. These ecosystems thrive in the dark by capturing chemical energy from the hot vents. The bacteria that form the basis of these deep ocean food chains are called *lithotrophs*, literally rock-eaters. Like plants, they extract carbon dioxide from seawater and string the atoms together to make sugars; but, unlike plants, they use minerals (principally hydrogen sulfide) spewed out of the volcanic vents as a source of energy. The bacteria *eat* hydrogen sulfide; everything else eats the bacteria.

FIG 2.1   Giant tubeworms inhabiting a submarine hydrothermal vent system.
© Woods Hole Oceanographic Institution.

Christian Lascu and Serban Sarbu discovered another lightless eco-system in a limestone cave in southern Romania. The cave appears to have been isolated from the surface for five million years; yet Lascu and Sarbu found transparent crabs, blind spiders and water scorpions crawling through its dark, damp interior. Microbial mats that cover the surface of a ground-water lake and the limestone walls of the cave, nourish the whole ecosystem. The bacteria appear to be able to extract carbon from limestone (calcium carbonate), using energy derived from the oxidation of hydrogen sulfide dissolved in the ground water.

FIRE AND BRIMSTONE

The Christian Hell is an inhospitable place: 'and he shall be tormented with fire and brimstone' (*Revelations*, 14:11). The most vociferous hellfire preachers conjure up images of fiery mountains, scorching deserts and bubbling pools of brimstone to roast the souls of mortals deserving eternal damnation. Yet, harsh though such environments might appear, they would in fact provide quite comfortable habitats for many (perfectly virtuous) living creatures.

Brimstone is an archaic name for sulfur, which is found in meteorites, hot springs and sprayed out of active volcanoes. It is often visible as pale yellow streaks decorating volcanic slopes. The element itself is relatively harmless. It gets its infernal reputation from its ability to float on water and burn, releasing poisonous fumes of sulfur dioxide. Many of its other compounds are also noxious. The reduced (reduction is the opposite of oxidation and often involves the addition of hydrogen atoms to an element or compound) compound of sulfur, hydrogen sulfide, is a foul smelling and poisonous gas (the smelly gas generated by the stink-bombs so beloved of schoolchildren). Sulfuric acid is one of the most corrosive acids. Yet sulfur is essential for life. Proteins and fats are particularly rich in sulfur. Our bodies contain about two-hundred grams of sulfur. The source of all this sulfur is bacteria, able to eat or breathe both sulfur and its noxious compounds.

We have already met hydrogen sulfide-eating bacteria at the depths of the ocean; but this metabolic capability is widespread. Sulfur-eating bacteria, such as *Thiobacillus*, live in soil and in fresh and saltwater; and use sulfur and hydrogen sulfide as an energy source to fix carbon dioxide and generate biomass. Other sulfur bacteria, such as *Desulfobacter*, live in brackish water and animal intestines and use oxidized forms of sulfur (sulfate) as we use oxygen – to breathe (they exhale hydrogen sulfide). Other bacteria such as *Chromatium* use hydrogen sulfide as a hydrogen source for photosynthesis, depositing the leftover sulfur granules within their cells. The combination of differing survival skills of various sulfur bacteria allows it to be cycled through the entire ecosystem of the Earth. On a smaller scale, miniature sulfur cycles take place within some warm fresh-water lakes fed by sulfur-rich steams. In the *sulphuretas* of Libya and Japan, it is cycled between its oxidized and reduced forms and in the process, elemental sulfur accumulates in the lake and is harvested commercially. So, far from being an instrument of infernal torture, pools of brimstone can be healthy and productive environments for many microbes.

Many of the bacteria that metabolize sulfur also generate sulfuric acid as a by-product. Although science fiction films such as *Alien* (1979) and its many sequels featured monsters with acid for blood, it is microbes, rather than monsters, which are most tolerant to acid. Acidity is measured in units of pH: pH7 is neutral, below pH7 is acid and from pH7–pH14 is alkaline. Our cells function within a fairly narrow pH range, from about pH7.5 to 8.5: very slightly alkaline. Blood contains a bicarbonate-based buffering system that maintains its pH within this range. These stores can however be depleted during illness, such as severe diarrhoea, resulting in the drifting of body fluids outside their normal pH range, causing metabolic acidosis or alkalosis. The consequences can be disastrous, leading to tissue damage, shock and death.

Other animals are much more tolerant of acid. Acid rain from the burning of fossil fuels has caused the acidification of many lakes in Europe and the USA. Some fish can survive in lakes with water as low as pH4 but if it becomes more acidic, all fish die. Yet these acid lakes still harbour many invertebrates and microbes. Algae, fungi and bacteria are able to tolerate the highest levels of acidity, down to about pH0. We have met some of these microbes already – the sulfur-oxidizing bacteria found in hydrothermal vent systems which excrete hot sulfuric acid. Many of these bacteria can grow at concentrations and temperatures of sulfuric acid that would dissolve metals. Even our own bodies harbour acid-tolerant microbes. Our stomach contents have a pH of 1–2. The acid not only helps to digest our food but kills microbial pathogens such as salmonella, which normally have to be ingested in huge numbers (generally more than a million) to cause disease. However, a few microbes do survive within our stomach's acidity, most notably the spiral bacterium *Helicobacter pylori*, colonize the stomach lining and cause ulcers.

A remarkable feature of acid-tolerant microbes is that the insides of their cells are not particularly acidic – about pH6. Acidity is a measure of hydrogen ion concentration. It is a logarithmic scale so that pH zero has one million times the concentration of hydrogen ions as pH6. Somehow the bacteria are able to maintain a million-fold concentration difference of protons (remember that a hydrogen ion, H^+, is just a proton) across their cell membranes. It is not entirely clear how the bacteria achieve this feat; presumably either by excluding protons from their cells or by possessing a very efficient proton pump to pump them out.

The extreme alkaline end of the pH scale (10–14) is also harmful to most animals and plants. Strong alkaline solutions such as caustic soda dissolve cell membranes and destroy cells. Many plants and microbes are however fairly tolerant of soils that may have pH values up to about 10. Environments more alkaline than pH10 are rare on this planet. The only stable systems are soda lakes fed by bicarbonate-rich natural springs. The pH of these lakes may be as high as 11.5, yet they are often rich in microbial life.

High concentrations of salt are toxic to most living organisms; as attested by salt-curing to prevent microbial growth and preserve meat and fish. When cells are suspended in salt, their internal water is *sucked out* of their cells by *osmosis*, which dehydrates and eventually kills cells. There are however many natural saline environments on Earth. The sea, with a salt concentration of about three per cent, is toxic to most land animals and plants but is of course haven to marine creatures. The Dead Sea is a twenty-eight per cent solution of salts, nearly ten times the salinity of sea water. Yet the Dead Sea is far from dead. Although no fish swim in its waters, it contains algae and a rich microbial flora. One of these microbes, called *Halobacterium*, produces a purple pigment, *bacteriorhodopsin* that is able to harvest light energy and is the only non-chlorophyll based natural light-harvesting system that we know of. Halobacteria are so salt tolerant that they can survive intact inside salt crystals. Salt-loving bacteria employ two principal mechanisms to survive the osmotic pressures of their saline habitats. The first is simply to accumulate lots of salt (usually potassium chloride) within their own cells. The second strategy is to synthesize large quantities of small organic molecules (like glycerol) inside their cells, which counteract the pull of the external salt.

The Gulf War left devastation in the Persian Gulf. Burning oil wells belched noxious black smoke and leaked millions of tons of crude oil into the surrounding land. It was an environmental disaster that many predicted would take centuries to mend. Yet only a few years later, wild flowers returned to the oil well sites. The key to the rapid recovery was the presence of oil-eating microbes in the soil. Many microbes can tolerate or even feed on chemicals poisonous to plants and animals. The soils surrounding the oil wells were probably already rich in these microbes before but thrived in the oil-polluted soil left by the war. The microbes fed on the crude oil, degrading it into less non-toxic chemicals. Microbes are able to feed on a wide range of chemicals poisonous to many other creatures, such as benzene, toluene, cyclohexane and kerosene.

LIFE WITHOUT AIR

Living and breathing are, for humans, inextricably linked. We talk of 'the breath of life'. Animals need oxygen to live. Yet there are many living organisms for which the breath of life is poisonous. Microbes, known as anaerobes, do not breathe air: indeed many are instantly killed on exposure to oxygen. This sensitivity makes anaerobes difficult to study and so their prevalence has not been appreciated until fairly recently. It may come as a surprise to you to know that more than eighty per cent of your faeces is made up of anaerobic bacteria. Very little air penetrates our lower bowels, making it an ideal environment for the proliferation of these microbes. The vast majority of these gut bacteria are entirely harmless colonizers of our intestinal tract or even beneficial; but they may occasionally cause problems (mostly abscesses and ulcers) particularly if introduced into the rest of the body by wounds or surgery. Anaerobic microbes are also widespread in the environment. They are found in the soil and in fresh and seawater, particularly in the air-depleted sediments at the bottom of lakes, rivers, seas and oceans.

So how do anaerobes live without oxygen? It's easy – it's living *with* oxygen that is the difficult feat. Toxic to all living organisms, oxygen is highly reactive – reacting with tissue to generate even more unpleasant chemicals such as hydrogen peroxide (used to bleach hair) and molecules known as *free radicals*. Air-breathing organisms have an armoury of protective enzymes to remove and destroy these toxic chemicals. Strict anaerobes lack these protective enzymes and are killed by oxygen.

So why do we go to so much trouble to breathe air when one of its chief components, oxygen, is so toxic? We use it to *burn* our food in the process known as respiration. Chapter 5 will examine respiration more closely, but briefly: electrons are harvested from our food and rolled down a kind of energy cascade to oxygen. The difference in energy (high-energy electrons from food to low-energy electrons in oxygen) is captured, providing energy for the cell. Oxygen-based respiration is a very efficient means of extracting maximum energy from food and has thus superseded the anaerobic metabolism in most higher organisms.

Some anaerobes *burn* their food in respiration, but not with oxygen. Many minerals (such as sulfate or nitrate) serve as low-energy electron dumps for their respiratory cascade. You may well have noticed that the sand alongside estuary waters is often black and smelly. The bad smell is hydrogen sulfide and the blackness is due to the presence of iron sulfide, both products of anaerobic bacteria's respiration in these estuarine waters. In fact, a wide range of minerals can be utilized by bacteria for respiration; some bacteria can even *breathe* iron.

Still other anaerobes do not actually respire at all but derive their energy from chopping up their food molecules into small pieces, usually into simple acids or alcohol, in a process known as fermentation. Fermented foods and beverages like wine, beer, sauerkraut, cheese and even coffee, depend upon the actions of these busy microbes. Anaerobes are of enormous ecological importance since they are often responsible for the final decay of organic matter. A giant compost heap would have long since enveloped the whole planet were it not for these bacteria. Cows and other ruminants have learned to harness the powers of these microbes. Their stomachs house an internal compost heap of plant material decomposing through the activity of billions of fermentative bacteria. One of the by-products of their fermentation is the greenhouse gas, methane. The huge quantities of the gas flatulently emitted by (the bacteria inside) domesticated ruminants is thought to contribute significantly to the greenhouse effect.

Far from being the breath of life, life carries on very well in oxygen's complete absence. This must of course be so, since (as covered further in Chapter 4) life emerged on this planet in an atmosphere completely devoid of oxygen. It was only after photosynthetic plants and microbes began to pour oxygen into the earth's atmosphere that aerobic life became possible on Earth.

JOURNEY TO THE CENTRE OF THE EARTH

Descend into the crater of Yocul of Sneffels, which the shade of Scartaris caresses, before the kalends of July, audacious traveller, and you will reach the centre of the earth. I did it.

ARNE SAKNUSSEMM

These were the instructions that, in Jules Verne's famous tale, led Professor Hardwigg and his companions to descend into 'the great volcano of Sneffels', following in the footsteps of the intrepid Icelander. Miles below the surface, they crossed a subterranean sea to discover a fabulous world of gigantic mushrooms, giant trees and extinct animals; flora and fauna from a world buried for millions of years. Scientist are now following in Professor Hardwigg's fictional footsteps to discover a world – though not as fabulous as that created by Jules Verne – that harbours many strange and remarkable creatures.

It was generally believed that terrestrial ecosystems extended just a few metres below either the land surface or the ocean bottom. Deeper than this living organisms were thought to peter out, as nutrients became scarce. However, oil exploration drilling in the 1970s started turning up microbes from deep inside the Earth. It was initially believed that the bacteria represented surface contamination of the drilling equipment. This view changed dramatically in 1987 when the Department of Energy (DOE) in the USA started to explore the storing of nuclear waste below ground. To investigate the stability of potential sites, they drilled three deep holes into the sedimentary rock beneath Savannah in South Carolina and extracted cores from as deep as five hundred metres. To their astonishment, even the deepest cores contained abundant microbial flora with more than four hundred species of bacteria. Similar cores drilled into sedimentary rock seven hundred and fifty metres beneath the ocean have yielded similar numbers of bacteria. Although bad news for the DOE (who want to store their nuclear waste in sterile environments), it has provided microbiologists with a further habitat to explore. The deepest hole so far, a three-and-a-half-kilometre-deep South African gold mine has yielded rock-eating bacteria that can acquire energy from iron, manganese, sulfur, cobalt and possibly even gold.

Remarkably, the deep drilling has not yet hit any level where life peters out. In fact, in some studies the bacteria are more numerous the deeper the drilling. Most of these bacteria are thought to eat the buried organic matter trapped in the rock when the sediments were laid down millions of years ago. However, abundant bacteria have also been found to inhabit deep water-filled cracks of buried volcanic rock where little or no organic material has percolated down from the surface far above. In 1995, another DOE project drilled one thousand five hundred metres deep into basalt beneath the Columbia River valley in Washington State. The bacteria found were mostly *methanogens*, able to use hydrogen as an energy source to make methane, which they incorporate into their tissue. Methanogens are also common on the Earth's surface. The intestinal tracts of animals, particularly ruminants, are full of methanogens; as are boggy waters where the methane they generate may spontaneously ignite, causing the ghostly will-o'-the-wisp flames that dance over the water's surface.

Thriving microbial ecosystems may also be found below permanently covered ice-sheets. I have already mentioned the ice-covered Dry Valley Lakes of Antarctica as ecosystems totally isolated from the surface. A vast ice-buried habitat of Antarctica remains to be explored. Lake Vostok is a liquid-water lake, two hundred kilometres long, with an average depth of one hundred and twenty-five metres, which lies two miles beneath the Antarctic ice sheet. The lake was only discovered in 1974 and, as far as we know, has been buried for at least a million years. There are plans to drill down into the lake and sample its ancient waters. The danger is that the sampling will contaminate its pristine waters, so a drilling programme in 1996 stopped just one hundred and fifty metres above the lake surface whilst scientists consider the best way to proceed.

LIFE WITHOUT WATER?

We have already encountered some of Earth's driest places, since they are also the hottest (deserts) and coldest (Antarctic Dry Valleys). As we have seen, many organisms, such as lichen, manage to survive drought conditions, but do so in a dormant state awaiting the return of moisture from melting ice, rain, fog or dew. The key to long-term survival appears to be a carefully controlled desiccation – removal of water under conditions avoiding damaging the cell. A commonly used technique for long-term storage of microbes and plant seeds is freeze-drying, in which water is evaporated whilst the cells remain frozen to minimize cell damage. Plants use a similar strategy to make drought-resistant seeds. The seeds undergo a process of controlled desiccation, in which water is replaced by a sugary liquid hardening to vitrify the seed.

Animals and vegetative plants do not tolerate drought. There are however a few plants, known as resurrection plants, which can survive conditions that reduce their moisture content to less than ten per cent. The palm-like fern, *Actiniopteris semiflabellata*, adorns exposed rock faces throughout East Africa. In times of drought, the plant dries to a crisp brownish-grey discolouration on the rocks; yet, when the next rains arrive, the dehydrated leaves absorb the water, resuming growth. Resurrection plants use a variety of mechanisms to resist the damaging effects of drought. Water is sometimes replaced by sucrose, which encases their cells in a glassy fluid. In other plants, a group of proteins, called *dehydrins*, appear to protect delicate cellular structures during desiccation.

Survival is, however, not active life. Seeds and drought-resistant plants are never active. Although, paradoxically, removal of water appears to be essential for long-term survival of dormant forms, it remains essential for active life.

SO WHAT ARE THE LIMITS?

The spacecraft's exploration of life would thus have discovered its extraordinary versatility. Life on Earth, particularly microbial life, knows few limits. The minimal ingredients appear to be simply sources of carbon, nitrogen, oxygen and hydrogen plus a few minerals – elements abundant both on this planet and elsewhere. It doesn't seem to matter too much how these are supplied; living organisms, particularly bacteria, are able to utilize sources as diverse as air, rock or vegetation. Active life also needs energy but organisms can capture either light energy or a multitude of chemical forms of energy.

Liquid water appears to be the chief limiting factor to life on Earth. Living organisms have a very limited ability to manipulate the freezing or boiling point of water: when the exterior temperature exceeds the limits of this ability, active life ceases. The most barren places on Earth are generally the driest. The relative sterility of the Antarctic Dry Valleys epitomizes the requirement for water but our own homes strikingly illustrate the same principle. Home maintenance is essentially a battle against moisture. We repair roofs and windows, paint surfaces with water-repelling chemicals and make endless trips to the DIY store in our battle to exclude moisture and promote desert conditions inside our houses. If we neglected this then microbes and moulds would quickly invade and undermine our homes.

Why water in its liquid state is so essential to life is a question we will return to in Chapter Five. On Earth, so long as liquid water is available, then life is also possible. Microbial life thrives in a diverse array of (watery) chemical environments from hot to cold, acid to alkaline and every other extreme of chemistry available on this planet. The source chemicals used to make up living cells are incorporated by a wide variety of chemical pathways and transformed inside living cells by a host of diverse metabolic pathways. There appears to be no common core of metabolic chemistry that drives all living cells. The diversity of the chemistry that underpins living cells in different creatures is surprising given the usual *chemical* explanation for the phenomenon of life: that it is a highly complex self-organizing chemical reaction (we shall be examining this in Chapter Six). If life is a merely a complex chemical reaction then we must explain how such a wide variety of chemical processes generates essentially the same phenomenon: life. This indicates to me that we need to look further than standard chemistry to discover the essential quality of life.

Another curious feature of life apparent from our brief exploration is its invasiveness. The inanimate world is characterized by a flowing down of energy: water flows down a hillside; electrons flow down to lower energy states and complex molecules break down to simpler ones. Yet living organisms will climb to fill any vacant niche. When in 1883, a volcano exploded on Krakatau, it obliterated two thirds of the island, leaving the remainder an inhospitable nutrient-depleted dustbowl. Yet some microbes (mostly cyanobacteria) can survive on a diet of volcanic ash and rapidly colonized the island. The growth of these microbes provided the nutrients to support further colonization, and a few decades later there was abundant plant-life and even a few small animals. Over the course of evolutionary time, life has relentlessly thrust itself in all possible directions to fill the planet's every available niche. This aspect of life, its ability to direct itself forward, is hard to account for in terms of the perpetual running down that dominates inorganic chemistry. We are again pointed towards the realization that life represents a very different kind of chemistry from the reactions that drive the inanimate world.

EXTRATERRESTRIAL LIFE?

After finding the limits of life on Earth, the spacecraft would surely explore our solar system to discover whether life is limited to its third planet. We live on a planet orbiting one star in a galaxy of one hundred billion stars in a universe of a billion galaxies. Is it conceivable that we are alone?

Science fiction writers have dreamed of all sorts of non carbon-based life forms but, although entertaining, none is convincing. Life is a complex business that requires complex chemistry. As far as we know, carbon is unique in its ability to form the wide range of compounds necessary for the emergence and evolution of any life form. Let us, thus, concentrate our thoughts on carbon-based life. Carbon is relatively abundant in today's universe. Our sun is about 0.3 per cent carbon. It is found with varying abundance on the planets and comets of our solar system in the form of carbon dioxide, methane and more complex hydrocarbons – all compounds used as carbon sources on Earth.

The next ingredient for life, hydrogen, is the universe's most abundant element. Oxygen and nitrogen are more sporadically distributed but are nevertheless relatively abundant. Minerals are scattered throughout the galaxy. Energy sources are certainly widespread; we have only to look at the stars to see billions of them. Alternative chemical sources of energy such as volcanism and geothermal energy also exist within our own solar system.

The key requirement that would limit extraterrestrial life is likely to mirror that which limits life on Earth: the presence of liquid water. Wherever liquid water is present on Earth, life is also found. It seems reasonable to extend that principle beyond our planet and predict that wherever stable bodies of liquid water co-exist with sources of carbon, nitrogen, hydrogen and oxygen, then life will also be found.

How abundant is liquid water in the universe? Water itself is not a problem. It is found on other planets of our solar system. It is abundant in comets and has been detected around extrasolar stars. The important question is rather: is water present as a liquid? The range of temperatures even within our own solar system is enormous: from the billions of degrees found in the sun's interior to only a few degrees above absolute zero in the outer solar system. Clearly, the upper end of the temperature scale is incompatible with even the existence of water as the molecule would disintegrate into its component atoms. Going down the temperature scale, there are thousands of (hot) degrees where water exists as a gas. A tiny window exists (just about 100°C at terrestrial atmospheric pressure) where water exists as a liquid. Below zero there are 273 degrees between freezing and absolute zero where water is present as solid ice. The feasibility of extraterrestrial life reduces to the bare question: does this liquid water window exist on other planets?

The closest planet to our sun is Mercury. It has the widest range of temperature for any planet in our solar system. At night, the temperature on the surface of the planet drops to −183°C and during the day it rockets above 300°C. The planet has little atmosphere and no detectable water, so it is a highly unlikely supporter of life.

The second planet from the sun, Venus, seems, initially, a much better prospect. Venus has a thick atmosphere consisting largely of carbon dioxide but with both nitrogen and water vapour also present. The thick atmosphere obscures all detail of the planet, allowing nineteenth-century writers and illustrators to imagine a tropical paradise inhabited by carefree, amorous Venusians. However when probes were sent to explore Venus in the 1960s they brought back images of a reddish brown rock-strewn desert beneath an orange sky. With surface temperatures a baking 480°C − far too hot for the existence of liquid water − and thick clouds of hot sulfuric acid that rain onto the terrain below, Venus is far more like Hell than Paradise.

Conditions have not always been so harsh on Venus. There is evidence that the planet once had deep-water oceans similar to Earth's. But high levels of carbon dioxide in the atmosphere set up a runaway greenhouse gas effect, trapping the solar heat, drastically raising the surface temperature and evaporating the oceans. Venus serves as a terrifying reminder of the dangers of ignoring the warnings of environmental catastrophe on our own planet.

The third planet from the sun and its inhabitants is the subject of the remainder of this book so let us pass quickly on to the fourth planet. Mars and Martians are of course synonymous with popular notions of extraterrestrial life. In 1877, the Italian astronomer, Schiaparelli, drew detailed maps of the planet and identified linear features on the surface of Mars which he called *canali*, channels. The word was incorrectly translated into English as canals and, although these features were later found to be optical illusions, tales of Martian civilizations building complex irrigation systems to distribute their dwindling water supplies captured the popular imagination. The first detailed images of the surface taken by the Mariner probes were thus a big disappointment to Martian-watchers. There were no civilizations, no canals – and not a drop of water.

Though we know that there are no canal-building Martians on Mars, the planet remains one of the most promising candidates for extraterrestrial life. The atmosphere has plenty of carbon dioxide, together with nitrogen and small quantities of water vapour. The ingredients of life are there but the planet is cold. The average surface temperature is $-53°C$: too cold for liquid water to exist on its surface. Yet liquid water did once flow on Mars. Networks of branching valleys with fine tributaries look remarkably similar to the Earth's river valleys. Surface features record what appears to be catastrophic flooding by rivers more than one hundred times bigger than the Mississippi. The river valleys, lake beds and flood plains are all dry now but they record a warmer and wetter period in Martian history. It is thought that this warm wet period ended about three and a half billion years ago, but that might have left just enough time for life to evolve (like Earth, Mars formed about four billion years ago). Bacteria were already well established on Earth three and a half billion years ago.

If microbes once flourished in Martian seas, they must have gone through a catastrophic crisis when the planet's surface dried up. The last stand of these microscopic Martians might have come when the dwindling seas, rivers and lakes were freeze-dried in the thinning atmosphere. But perhaps there are still outposts of life on Mars. Though the planet's surface is now dry, its crust is estimated to hold a layer of water-ice five hundred metres thick. This permafrost layer would not be much different to that of the Dry Valleys of Antarctica, which does harbour life. Could Martian bugs – refugees from the ancient seas – survive still in the frozen subsurface? At present, we simply don't know. The key feature allowing life to survive in Antarctica are the brief warm summer spells when the ice melts, releasing liquid water. Mars lacks a warm summer but it does have other sources of heat. Martian volcanoes like the massive *Olympus Mons*, five hundred and fifty kilometres across and twenty-five kilometres high, are potential sources of geothermal energy. The heat from volcanic eruptions must have melted huge quantities of the subsurface ice and probably caused the catastrophic flooding episodes recorded on the Martian terrain. Whether sufficient water remained liquid long enough to sustain life is, of course, very uncertain.

Geothermal energy may still be active under Mars' surface. Mars almost certainly has a hot core like Earth's. Although the surface is frozen, it is likely that temperature increases with increasing depth. There must exist a subsurface temperature window, hot enough to melt ice but not too hot to vaporize it. On Earth, microbes live in the deep subsurface where liquid water is present and may have survived there for millions of years. Similar conditions under the surface of Mars may yet harbour Martian microbes.

The possibility of life on Mars recently hit the headline with the publication of images of supposed fossilized microbes (see figure overleaf) buried inside a Martian meteorite. The brick-shaped meteorite, known as ALH 84001 weighed nearly two kilos and was collected in the Allan Hills area of Antarctica. The rock was a *basalt* which had solidified from volcanic lava about four and a half billion years ago. But no earthly volcano spewed out ALH 84001. Analysis of gases trapped within the rock identified it as a small piece of Mars. Around about three and a half to four billion years ago, carbonate minerals were deposited in the rock, possibly precipitated from groundwater seeping through the Martian

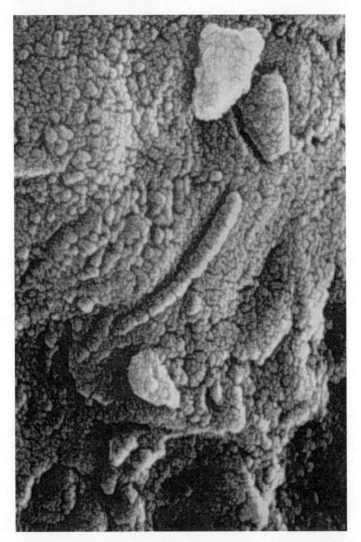

FIG 2.2   Electron micrograph of Martian meteorite ALH 84001.
Spherical and worm-like structures have been proposed to represent
the fossilized remains of ancient Martian microbes. Johnson Space
Center, accession number s9612609.

surface. The rock remained on Mars for the next three billion years and
would still be there if a comet or asteroid had not crashed into Mars
about sixteen million years ago and ejected the rock into space. After
spending an uneventful few million years drifting through space, it was

captured by the Earth's gravitational pull and fell down on one of Antarctica's blue-ice fields about eleven thousand years ago. In 1984, an ANSMET (Antarctic Search for Meteorites) team of scientists found and collected the rock, dubbing it ALH 84001.

The meteorite rock was packed in dry ice and shipped to the Antarctic Meteorite Laboratory at Johnson Space Center in Houston, Texas. There, it was catalogued and classified as a 'common' asteroidal meteorite. Its Martian origin was not discovered until 1993 when scientists took a closer look. Not only identifying it as coming from Mars (only the twelfth known Martian meteorite), researchers sectioned and examined the rock under the electron microscope and discovered globules of carbonate minerals and structures that looked remarkably like microbial fossils.

The Martian microfossils may look like bacterial fossils but, as any geologist will tell you, there are many natural rock formations that resemble fossils. The research team headed by Dr David McKay of NASA's Johnson Space Center, supported their claim by also reporting chemical evidence of past life in the rocks, in the form of chemicals known as polycyclic aromatic hydrocarbons. However, if the structures do represent the remnants of bacteria, they are very significantly different from modern bacteria. Bacteria alive today are in the micrometer size range. The microbes that cause *trachoma* – an infectious disease that leads to blindness – are among the smallest. These *chlamydia*, have spherical cells measuring only a third of a micrometer (a millionth of a metre) in diameter. Yet the Martian 'microbes' are in the nanometre (a billionth of a metre) size range, and are usually only about 10 nanometres long. The cells could have had only a very tiny volume, about one millionth to one thousandth of the volume of a typical bacterium. Clearly, they couldn't have held much material inside.

Yet, *nanobacteria* may also be found on Earth. Examination of the deep subsurface rocks recovered from Columbia River basin project, has revealed structures that look like nanobacteria; although their biological origin has not yet been confirmed. Robert Folk of Texas University claims to find nanobacteria in material from tapwater to tooth enamel.[4] There have even been reports of nanobacteria recovered from human blood. Perhaps nanobacteria represent an earlier phase in the evolution of life. As we will be discussing in Chapter Four, it is highly unlikely that cells as big and complex as modern bacteria could have been the earliest life forms on Earth. The proposed nanobacterial structures formed on Mars at about the same time as life originated on Earth. If life was also in its infancy on Mars, then the nanobacteria fossils may be relics of the earliest life.[5]

Studying Martian life by examining rocks blown off its surface clearly has its limitations. The best way to look for life on Mars is to go there and examine the rocks directly. The late 1970s Viking mission to Mars did just that and hunted for evidence of life on the surface. Although it did discover a peculiar chemistry that mimicked biochemical activity, it is generally thought that the findings were negative. However Viking only sampled surface soils and it is likely that to find life on Mars you would have to dig deep. The current posse of Mars probes, including the Mars Pathfinder Mission's indomitable rover vehicle, *Sojourner*, do not have any microbe-hunting experiments. But the interest generated by the recent Mars meteorite story prompted President Clinton to promise the 'full intellectual power and technological prowess of the US behind the search for further evidence of life on Mars'. Let's hope that future Mars missions have drills on board.

Beyond Mars, we come to the giant gas planets – Jupiter, Saturn, Uranus and Neptune. These have the necessary ingredients for life: hydrogen, methane (a carbon source), ammonia (a nitrogen source), and water. But they are very cold. The temperature on the cloud tops of Jupiter is a chilly −153°C. Vast oceans of liquid hydrogen may lie beneath the clouds of the giant planets with solid cores probably ten to twenty times as massive as Earth. It is possible that liquid water may exist at some altitudes within their atmospheres. In a fanciful moment, the late Carl Sagan proposed that Jovian life might take the form of floating bag creatures that drift through the Jovian atmosphere. The Jovians would however have to endure a racy existence, driven by the two hundred and fifty miles an hour winds that blow through the upper atmosphere. All in all, the giant planets look unlikely habitats.

The outermost planet, Pluto, is smaller than the moon and has a surface temperature of about −236°C. It looks the least likely place to find life in the solar system. More hopeful sites are on some of the moons of the giant planets. One of Saturn's moons, Titan, has a thick atmosphere with water and traces of at least a dozen carbon-based compounds, including methane, ethane, hydrogen cyanide and carbon dioxide. The mixture is similar to the atmosphere many scientists believe existed early in Earth's history, when life first emerged. But the temperature on Titan is a chilly −180°C, far too low for liquid water. Its similarity to the early Earth has led to its description as *Earth in the deep-freeze*.

The young contender of exobiology candidates is the Jovian moon, Europa. About the size of our moon and with a surface temperature of −145°C, Europa does not at first look a likely candidate. However, when the Galileo spacecraft sent back detailed images of the moon's surface, it looked familiar. In fact the pictures could have been taken from above the Antarctic ice packs. Europa is entirely covered by a thick sheet of ice. The ice layer is probably about one hundred and fifty kilometres thick but evidence is accumulating that it is not all ice. Close-up shots reveal a cracked and broken surface and structures which look remarkably like icebergs (see the figure overleaf). Something must be causing the ice to crack and break and the betting is that a liquid water ocean is churning up the surface ice, exactly like pack ice on Earth. Recent optical data from Galileo has detected mineral salts on the ice surface, probably the dried up remnants of briny seawater extruded onto the surface.

Scientists speculate that geothermal or tidal energy may be the heat source that has melted the putative ocean beneath Europa's ice. Perhaps hydrothermal vents similar to those discovered by *Alvin* exist on Europa, spewing out hot mineral-rich water into the ice-locked ocean. Galileo's instruments have detected complex carbon-based compounds on Europa's sister moons, Callisto and Ganymede, making it highly likely that similar compounds are present in Europa's seas.

The ingredients are all there. Europa almost certainly has a liquid water ocean with sources of carbon, nitrogen, minerals and a geothermal energy source. Similar conditions on Earth support complex ecosystems. Do Europa*eans* swim beneath the ice of Europa? On the principle that there is nothing special about Earth, my prediction would be (a hopeful) yes. Many scientists consider the imminent exploration of the terrestrial Lake Vostok as a rehearsal for a robotic dive beneath Europa's ice, early in the next century. Perhaps the new millennium will be marked by our first contact with alien life.

And beyond the solar system, is there life among the billions of stars in our galaxy? Using the same approach applied to the solar system, we would predict life on planets possessing the necessary ingredients of

FIG 2.3   Ice rafts on Europa that probably float on a vast water
ocean. National Space Science Data Center Photo Gallery, NASA
Photo Number P-48526.

carbon, hydrogen, nitrogen, oxygen, minerals and liquid water. These
elements are certainly common throughout the galaxy so it is unlikely
that life is limited by a lack of raw materials. The more difficult problem
is to assess whether planets exist with liquid water. Until recently, nobody
knew whether extra-solar planets existed. This has changed dramatically
in the last few years with the discovery of many planetary systems around
distant stars. The planets are usually detected by the periodic wobbling
of a star, betraying the presence of a hidden companion object. So far

the detectors can only pick up giant planets, about the size of Jupiter or even bigger. They are likely to be gas giants and therefore unlikely hosts (though they may have solid moons that could harbour life). About a dozen of these giant planets have now been detected, and many more are expected in the coming years. There is no reason to believe that the giant planets are alone. Earth-sized planets are also likely to be orbiting these distant stars. The optical signature of water has been detected in at least one putative planetary system.

Beyond our galaxy lie billions of other galaxies. I think it inconceivable that terrestrial conditions do not exist on many of the billions of planets probably orbiting those billions of stars. However, the gigantic distances that separate us from even our neighbouring galaxies (the Andromeda galaxy is a neighbour, but travelling at the speed of light it would still take two million years to reach it) ensure that such questions will, for a long, long time, remain entirely academic.

My guess, for what it's worth, is that life is common throughout the universe. Just as life is found on Earth wherever we find the necessary ingredients alongside liquid water, then extraterrestrial life will be found wherever those conditions coincide. Astronomical evidence seems to be tilting towards an expectation that this combination is not so special.

We must now come down from the stars to return to this book's central quest: to understand life on our own planet. Life's success here on Earth has been contingent upon its most important action: the ability to replicate. Reproduction, the biological imperative, is clearly the most important action that (most) living creatures perform, so it is here we will begin our exploration of the source of life's actions.

3

Life's Biggest Action

Like nearly everyone else on this planet, I have a keen interest in sex. However, I justify my curiosity by claiming a fascination with procreation, or, simply, how living creatures make copies of themselves. Biologically speaking, reproduction is our most important action. We wouldn't be here if our parents and all our ancestors hadn't managed to reproduce before death. Even the icy microbes that live below Antarctica's frozen lakes must occasionally split into two cells.

All reproduction involves making a copy of the parent in a new body. To make a new body, we (or at least females of our species) need food and energy. We obtain these by the variety of strategies already described. More importantly, our cells and bodies need to *know* what kind of body to make and how to make it. Despite the variety and diversity amongst living creatures in the materials that they use to build cells and bodies, they all utilize the same material for storing their building instructions. Which brings us to the thorny question that has occupied most young minds at one time or other.

'WHY DO I LOOK LIKE DADDY?'

A landmark in biology was reached when in 1953 James Watson and Francis Crick discovered the double-helical structure of DNA. Books, magazine articles, documentaries and T-shirts have all vouched for this event's significance, but why? Why is the double helical structure of DNA so important? It is certainly an elegant, beautiful structure, and its elucidation was a tribute both to the experimental skills of Rosalind Franklin and the genius of the Watson and Crick partnership. But many other biological structures were being worked out at about the same time. A few years before the double helix was known, the Nobel prize-winning chemist Linus Pauling discovered that proteins had helical domains which he called alpha helices. Yet how many T-shirts are printed with the alpha helix pattern? Why is the double helix so famous? What problem did it solve?

It solved the two fundamental problems of biology – how biological information is encoded and how it is inherited. Or, *why you look like your daddy* (or indeed, mummy, but children who know that babies are made inside their mothers tend to find this less mysterious). Thankfully, the prudish sex education I experienced as a child ('Girls and boys, read pages 210 to 230 in your biology book and I don't want any questions') is a thing of the past. Today most schoolchildren would just as non-chalantly draw a picture of a sperm cell as of a sperm whale. Yet this knowledge, that we all take for granted, has been hard-won.

We know little about prehistoric views on sex education, but the widespread worship of deities represented as both heavily pregnant females and phalluses demonstrates the important role sex and repro-duction must have played in ancient ritual. The myths of the primordial egg of creation reveal man's familiarity with at least the egg-laying repro-ductive strategies of birds, fishes, amphibians, lizards and snakes. It is less clear how well our prehistoric ancestors understood human reproduction. Many early myths reflect a belief that women were impregnated either by divine intervention or through some natural agency. The legends of King Conchobar of Ulster relate how his mother conceived – by swallow-ing two worms in a cup of water.

But although many mythical heroic figures were said to have been born of these immaculate conceptions, recognition of the male role in the procreation of lesser mortals is well attested in many ancient tales. Neolithic farmers were surely familiar with seeds, pollen and their role in plant reproduction. The Babylonian Hammurabi's code (from roughly 1750 BC) mentions the practice of hand-pollinating date palms. Man's visible semen became to be equated with seeds and considered to be the seed of human reproduction. Aristotle reinforced this by claiming the male contributed the character of 'form' to reproduction whilst the female role contributed merely unorganised 'matter', to be moulded as clay by the male seed.[1] Aristotle also believed in the principle of *epigenesis*, in which each organism begins life as a formless mass which grows and differentiates into the head, limbs, organs and eventually the entire body of the individual. The ethereal soul, rather than matter, was thought to guide the development of the body.

Little of substance was added to Aristotelian embryology until the rise of mechanistic philosophy in the seventeenth century. A belief in the influence of an immaterial soul was of course anathema to the rationalists who instead embraced the curious theory of *preformation*, whereby seeds or eggs were proposed to contain the miniaturized parts of the adult plants or animals. Although today these ideas may appear absurd, they at least provided a mechanism to account how the information encoding the form of an animal body or plant could be passed through a vessel as small as a seed or egg. The answer was simply to propose that the people or plants began life as complete beings small enough to fit inside.

A new twist was added in 1677 when the Dutch draper (and inventor of the microscope) Anthony van Leeuwenhoek (1632–1723) observed 'little animals of the sperm' in human semen. It was a student from Leyden, Johan Ham, who first used the microscope to observe sperm swimming vigorously through human semen. Johan's uncle took him to see Leeuwenhoek, who confirmed his observations. Leeuwenhoek declared these sperm the carriers of miniaturized humans, 'Man ... already furnished with all of his members', the real creators of new life.

The proponents of preformation fell into two camps. The *ovists* believed that it was the ovum (the female reproductive cell of animals – embryologists tend to reserve the term *egg* for the variety you might find on your breakfast table) that provided preformed individuals. Opposing them were the *spermists* (like Leeuwenhoek) who maintained that it was the sperm that were the seeds of the next generation. The more enthusiastic preformationists went so far as to claim that they discerned perfectly formed, tiny human bodies (*homunculi*) enclosed within human ova or sperm. However, a further complication was that the beings inside either eggs or sperm should themselves have perfectly formed ovaries or testis within which should be found eggs and sperms with their own (even tinier) preformed individuals inside. This process could go on *ad infinitum*, in an ever-diminishing series. This problem did not daunt the preformationists who claimed that the ancestral Eve held within her ovaries the forms of all the men and women that would ever live, each embedded inside the other like Russian dolls.

The role of both sperm and eggs in amphibian reproduction was finally demonstrated by the Italian Lazzaro Spallanzi (1729–1799), who fashioned tiny taffeta pants for frogs to prevent insemination during mating and showed that under these circumstances, the eggs did not generate tadpoles. Spallanzi later collected unfertilized eggs and sperm from his frustrated frogs and demonstrated that the eggs developed into tadpoles only after they had been mixed with sperm. The microscopist Johannes Müller (1801–58) went on to observe spermatozoa penetrating the ovum of animals. The increasing power of microscopy, together with studies on the development of plant and animal embryos in the nineteenth century, led inevitably to the demise of preformation. In its place, the Aristotelian principle of epigenesis, in which new parts develop from an undifferentiated embryo, re-emerged.

But the absence of tiny individuals to carry their form into the next generation left the problem: how was the information to make an adult body carried from one generation to the next? Indeed, how is it encoded in the first place? How does the undifferentiated chick embryo *know* that it must grow into a baby chick rather than a baby mouse or a baby tree? These were the great, unanswered questions of nineteenth-century biology. How is biological information encoded and how is it inherited?

One solution was reversion to the Platonic concept of ideal forms. In the realm of the perfect circle and the perfect sphere, there may also be the perfect form of a mouse, man or tree. This theory would, however, only work if species were themselves unchanging, reflecting the permanence of the proposed abstract realm beyond physical reality. In the eighteenth and early nineteenth centuries this did not present an obstacle since most naturalists believed in the continuity of species. The Swedish father of taxonomy, Carl Linnaeus (1707–1778) was first to systematically classify species of plants and animals and, like his contemporaries, he considered all species to be permanent, created by God; insisting, 'There is no such thing as a new species.'

EVOLUTION

French aristocrat and naturalist, George Louis Leclerc, Comte de Buffon was among the first to question the immutability of species. Buffon noticed the presence of apparently vestigial parts in some animals, such as the bones of useless lateral toes in the pig. This led him to propose that species did change by degeneration of disused parts. His assertion that mammals degenerated in size in the New World provoked Thomas Jefferson to have the skeleton of a seven-foot moose sent to Paris, to prove 'the immensity of many things in America'.

Jean Baptiste de Monet Lamarck (1744–1829) made a far more radical case – for *transmutation* of species. Lamarck came to science by a circuitous route. He was sent to a Jesuit school by his parents to train to be a priest but at sixteen his father died, leaving a small inheritance, just enough to buy a horse. Lamarck immediately left the priesthood and rode off on his new purchase to join the army fighting in Germany. After a brief (but reputedly glorious) military career, Lamarck moved to Paris to work as a bank clerk and in his spare time studied medicine, music and botany. His botanical publications impressed the influential Buffon who helped him to obtain a position as assistant in the botany department of the Jardin du Roi – the king's botanical gardens. The French Revolution, bringing about the demise of his patron, was more favourable to Lamarck himself and he was elevated to the Chair of Zoology (but not botany) by the National Convention. One did not lightly refuse the dictates of a French revolutionary council so Lamarck turned his attention from plants to invertebrates.

Lamarck made many contributions to biology – not least its name. His career in both botany and zoology convinced him that all living things should be studied as a whole and so he introduced the term 'Biology' (from *bios*, Greek for 'life') to encompass these studies. He was an accomplished taxonomist who was the first to separate spiders and crustaceans from insects. But his most famous contribution is his theory of evolution. Impressed by the similarity between many species of insects and other arthropods, Lamarck noted how it would take only minor modifications of form to change one species into another. He proposed that *evolution* had done exactly that and that modern species of plants and animals were descendants of earlier extinct species. The source of evolutionary change is what separates Lamarck from most latter-day evolutionists. Lamarck believed that characteristics acquired during an animal's lifetime could be inherited by its offspring: the *inheritance of acquired characteristics*. His most famous example is the giraffe, which through many generations stretching for the leaves in topmost branches is proposed to have passed the acquired characteristic of a longer (stretched) neck to latter generations. Lamarck's ideas were ridiculed at the time and have received a bad press ever since. Yet, they at least freed biology from the concept of the immutability of the species, setting the stage for the most famous theory of evolution.

In 1859, Charles Darwin published *The Origin of Species* and changed biology forever. A public thirsty for science eagerly awaited the book. All one thousand, two hundred and fifty copies of the first edition were sold on the first day of publication. The controversy and debate that ensued continues to reverberate today. What Darwin did in *The Origin of Species* and his later works, was to place man firmly in the material world, an animal like any other. To those who refused to countenance a place for themselves amongst our hairy cousins, it was (and, in some quarters, still is today) considered heresy.

Darwin did not of course invent the theory of evolution, but it was in his family. Darwin's grandfather, Erasmus Darwin (1731–1802) was an eminent physician and keen amateur naturalist who proposed all organisms had descended from a 'primal filament'. Yet Charles Darwin did not look likely, initially, to follow in his grandfather's illustrious footsteps. His father, Robert Darwin, often harangued the young man, declaring, 'You care for nothing but dogs and rat catching, and you will be a disgrace to yourself and your family.' This assessment looked increasingly probable as Charles dropped out of first, Medicine at Edinburgh and then Divinity at Cambridge. When, in 1831, the opportunity arose to serve as unpaid naturalist aboard the HMS *Beagle*, his father was at first opposed to the idea. It was only after the intervention of Darwin's uncle, Josiah Wedgwood, that Robert Darwin was persuaded to allow his son to board ship.

Perhaps the elder Darwin did have a point about his rat-catching son because although Charles Darwin published *The Voyage of the Beagle* and several geological works in the years after his return from his voyage, he sat on his theory of evolution for twenty years. But in 1858 a letter arrived from Alfred Russell Wallace with an essay entitled, 'On the tendency of varieties to depart indefinitely from the original type'. Horrified to find that Wallace had independently come to the same conclusions, Darwin's first reaction was to abandon his lifework, but after much soul-searching and advice, he was persuaded to make a joint presentation with Wallace at the Linnaean Society. Darwin then set about organizing the mass of material he had collected over the years and published an abstract of his work, *The Origin of Species by Means of Natural Selection or the Preservation of Favoured Races in the Struggle for Life*.

What was original to the theory of Darwin and Wallace was that they proposed a mechanism to drive evolution – natural selection. The idea owes much to Malthus' *Essay on Population* which argued that mankind will always produce more offspring than the available resources can support. The inevitable consequences, according to Malthus, are war, famine, poverty and disease, as increasing populations fight for limited resources. To Darwin, the same pressures amongst living organisms lead to natural selection or survival of the fittest. Only the fittest individuals will capture the resources needed to generate offspring.

The survival of the fittest is not enough in itself to bring about evolution because there is as yet no mechanism to generate change. The vital extra ingredient for natural selection is therefore *inherited* chance variation within a species. Darwin observed how all individuals within a species were very slightly different. Breeders of domesticated animals, such as dogs, have exploited these small differences to select animals with the desired traits, such as the shape of a face or the length of a tail. After centuries of breeding, quite distinct breeds, such as the poodle and the Great Dane, have been generated from the same ancestral wild dog. In the same way, nature could act as the selective force. These naturally occurring, randomly varying individuals within a species would compete for resources. With progeny always outstripping resources, only the fittest variants would survive to reproduce. *Natural selection* would tend to *breed* from those fittest individuals who possessed the most desirable traits. Over millennia, natural selection would bring about gradual change in plant and animal characteristics towards increasing fitness – *descent with modification* – evolution.

Publication of *The Origin of Species* aroused a great deal of criticism. But although the ensuing theological debate generated the greatest controversy (and still does), a very real scientific problem led Darwin to at least a partial abandonment of his original hypothesis. The difficulty arose with the commonly held belief at the time that sexual reproduction always led to the *blending* of characteristics. This blending would tend to wipe out variation – so that if poodles and Great Danes were allowed to mate freely, then their offspring would tend towards nondescript mongrels. The physicist H. C. Fleming Jenkin and many others held that no amount of selection for either the characteristics of poodles or Great Dane could result in their offspring being anything other than a blend of the two.

Without variation, natural selection had nothing to select. Darwin needed some mechanism to replenish the variation that was being washed out of populations by blending. His answer was to embrace the discredited Lamarckian theory of inheritance of acquired characteristics. In the 1868 edition of *The Origin of Species*, Darwin proposed that parts of the body threw off minute particles congregating in reproductive cells to be inherited by the offspring. These bits, or *gemmules*, were held to transmit characteristics *acquired* during an individual's lifetime and thereby provided a new source of variation to offset the loss of variation by blending. The gemmules theory had more than a whiff of desperation about it and did much to discredit Darwinism in the latter half of the nineteenth century.

GENES

Two years before the 1868 edition of *The Origin of Species*, an obscure monk working in an Augustinian monastery in Brünn had discovered and published the right answer to Darwin's problem. Born into a peasant family in Silesia, Gregor Mendel joined the monastic life to escape poverty. In the monastery garden, Mendel crossed pea plants and carefully recorded the inheritance of parental traits – such as round or wrinkled peas – in those hybrids he generated. He demonstrated that, contrary to expectations, discrete characteristics such as the shape of the pea seed *did not* blend in crosses. Instead they *bred true* – were unchanged when they appeared in subsequent generations – though sometimes skipping a generation. To account for his peas, Mendel proposed that organisms contain within them discrete factors, passed unaltered from one generation to the next. We now call Mendel's discrete factors, *genes*. This was the exactly the answer Darwin needed – genetic variation would not be lost during sexual reproduction but emerge, unscathed by its passage, in each generation.

Darwin died in 1882, completely unaware that the work that would rescue his seriously flagging theory was languishing in the Linnaean Society library, the very place his own theory had been triumphantly unveiled. Even more tragically, Mendel died in obscurity in 1884, his revolutionary work unknown or forgotten. It was not until 1900 that three botanists, Hugo de Vries, Carl Correns and Erik von Tschermak, independently rediscovered his experiments. Working on the inheritance of variation in plants, they were each finding evidence for discrete patterns of inheritance. Searching the literature for any related similar work, independently each came across brief references to Mendel's publications and immediately realised their significance. Mendel was posthumously recognized as the father of modern genetics. Mendelian genetics went on to revolutionize twentieth-century biology and medicine.

The early twentieth century saw the fusion of Darwinian evolutionary theory (his original rather than the Revised Version) and Mendelian genetics in what has come to be known as the *neo-Darwinian synthesis*. In 1901 De Vries published the first volume of his *Mutation Theory*, in which he proposed that evolution occurs by discrete steps or mutations that were rare chance modifications of Mendelian genes. These mutations were the source of the variation for Darwinian natural selection and evolution. Evolutionary theory was, at last, complete.

But what were genes? What were they made of? How were they inherited? How were they modified? In 1901 nobody had any idea. The first real clue did not come until 1945 when the bacteriologist Oswald Avery's experiments at New York's Rockefeller Institute, demonstrated that genetic characteristics could be transferred from one bacterial cell to another, simply by transferring a chemical called deoxyribonucleic acid, or DNA. Avery purified DNA from some bacterial cells which produced a capsule (a slimy protective layer made of strings of sugars that surrounds the bacterial cells). He found that if he added it to cells that didn't produce a capsule then some of them would be transformed into capsule-producing bacteria. It appeared that the genetic information, the gene for the capsule, was made of DNA and could be transferred from one bacterial cell to another simply by transferring the DNA chemical.

However, not everyone was convinced of the significance of Avery's demonstration that DNA encoded bacterial slime. DNA was considered an unlikely vehicle for heritable information. Different species were assumed to have different genes but DNA isolated from different species appeared identical. The prevailing opinion was that genes were made of the protein. It was easy to show that different species had different proteins. Proteins contaminated all preparations of DNA, so many scientist's believed that it was the contaminating proteins in Avery's experiments that had transferred the genetic information. When in 1944, the quantum physicist Erwin Schrödinger (of whom much more later) published his book, *What is Life?*, he went along with the prevailing *genes as proteins* hypothesis.

However, in the early 1950s Alfred Hershey and Martha Chase's experiments proved that genes were made of DNA. They demonstrated that when a virus infects a bacterial cell, it injects its DNA *but not its protein* into the host cell. After infection, the bacteria become transformed to make the bacteriophage proteins. So the genes encoding those proteins must have been injected with the bacteriophage DNA. Genes must be made of DNA.

DNA was quickly accepted as the genetic material, but it was unclear how genetic information was stored within it. The overall chemical composition of DNA was already known – it was composed of a simple sugar (deoxyribose), phosphate groups and roughly equal quantities of four types of *nucleic acids*, each made up of carbon, nitrogen and hydrogen atoms. But the profound problem remained – how do these chemicals store the information for the shape of your nose? This was answered by Watson and Crick's structure.

THE DOUBLE HELIX

The intertwined DNA molecule has become such a strong cultural icon today that it is hard to realise just how *unobvious* its structure was in the 1950s. Both laboratories racing for a solution initially got it wrong. Linus Pauling of the California Institute of Technology (CalTech) was perhaps the greatest chemist of the twentieth century. He had already discovered proteins contained helical regions so it was hardly surprising that he proposed a helical structure for DNA. However, he wrongly proposed a triple helical structure.

James Watson was an American scientist who came to Cambridge's Cavendish Laboratory to learn protein biochemistry. However, his real interest was DNA and at Cambridge he teamed up with the Englishman Francis Crick to solve the structure of that 'most golden of all molecules'.[2] Watson and Crick's first stab at a structure for DNA was (like Pauling's) also a triple-stranded helix. The pair rashly invited the UK experts on DNA, Maurice Wilkins and Rosalind Franklin, to the unveiling of their putative structure. Wilkins and Franklin had travelled from King's College, London to view the new model but felt their trip had been wasted when it took just a few minutes for Franklin to spot crucial flaws in the triple helix. After a hasty, tense lunch, the King's Group rushed off to catch their train home.

News of this débâcle soon reached Sir Lawrence Bragg, chief of the Cavendish laboratory and Watson and Crick were instructed to turn their attentions to less challenging molecules. The pair decided to continue surreptitiously with their model-building. Their approach used wire models representing the chemical groups to build (quite literally) a structure in three-dimensional space that, they hoped, would represent DNA's actual structure. But how could they know whether their structure was correct? The key was Rosalind Franklin's X-ray crystallography data – which was essentially an X-ray of the DNA molecule. The problem, according to Watson, was the difficulty they experienced getting a look at the data over the shoulder of the allegedly overly secretive Franklin. Yet between the lines of Watson's very readable account, *The Double*

Helix, you can read something of the cultural landscape that forged Franklin's diffidence. Watson's describes his female colleague:

> 'Though her features were strong, she was not unattractive and might have been quite stunning had she taken even a mild interest in her clothes. This she did not. There was never lipstick to contrast with her straight black hair, while at the age of thirty-one her dresses showed all the imagination of the English blue-stocking adolescents.'

There is unfortunately no record of Franklin's opinion of the good looks and dress sense of her male colleagues. Some inkling of her likely feelings may be garnered from Watson's account of a particularly frank exchange of views between himself and 'Rosy' which ended with her bearing down upon him with a glint of violent retribution in her eyes. Watson was saved from a justly deserved slap by the entry of Franklin's colleague, Maurice Wilkins.

With the help of Wilkins (though behind Franklin's back) Watson and Crick finally managed to get a good look at her latest X-ray pictures of DNA and were quick to recognize the telltale image of a helix. Several days of frantic model-building resulted in their triumphant unveiling of the double helical form of DNA. The first to share in the newly discovered 'secret of life' were the regulars of the Eagle, the pub just outside the Cavendish laboratories. But a paper was soon drafted to *Nature* which ended with the classic understatement: 'It has not escaped our notice that the specific pairing we have postulated immediately suggests a possible copying mechanism for the genetic material.' According to Crick, their intention was not to be coy but was born of Watson's fear that he might 'make an ass of himself' by saying too much too soon.

HOW DNA WORKS

The key to Watson and Crick's structure is the order and pairing of the nucleic acid bases. The backbone of each DNA molecule is a string (polymer) of deoxyribose sugars, linked by phosphate groups. Each sugar has a single base attached that can be one of four bases, guanine (G), cytosine (*C*), thymine (*T*) or adenine (*A*). Running down the length of a single DNA strand you can therefore read a linear sequence of bases, such as *ATCCGTACCTGAACATAACCGATT*... Codes were not unfamiliar in post-war England, particularly to Crick who during the war had worked as a scientist for the Admiralty. The linear sequence of bases looked like a code: the genetic code. Watson and Crick suggested that the sequence of bases codes for the structure of proteins. By the 1950s it was known that proteins performed nearly all the work of making living cells. In particular, enzymes which make everything else DNA, RNA, fats, sugars, polysaccharides – the complete living cell – are proteins. If DNA encoded the information to make proteins, and proteins made everything else, then the problem of how cells *know* what to make would be solved.

Over the next decade the code was cracked, confirming that DNA sequences did indeed code for proteins. Proteins are linear polymers of amino acids (another group of simple organic acids). There are twenty common amino acids that go into proteins but only four bases that go into DNA. There cannot therefore be a one-to-one coding between a DNA base and an amino acid. It was not long before experiments performed by Marshal Nirenberg, Gobind Khovana and Severo Ochoa established that a triplet of bases, called a *codon*, encodes each amino acid. The codon *GCC* for instance codes for the amino acid alanine, whilst *GGC* codes for glycine. A protein made of only 1,000 alanine amino acids would have a genetic code consisting of 1,000 codons (3,000 bases); each codon being the sequence *GCC*, generating a DNA sequence *GCC GCC GCC GCC GCC*... All natural proteins are far more complex than this and are encoded by a more complex code; but the principle is the same.

This was the answer to one of life's great puzzles: how biological information is encoded and stored inside living cells. The DNA sequence encodes the sequence of proteins and the proteins make everything else (even DNA itself – a curious example of *self-reference* which is one of the intriguing features of life). By directing the synthesis of proteins, the DNA molecule is able to orchestrate all the activities of the entire cell – and thus the entire body. This is how a dog cell *knows* how to make a dog, how an oak cell *knows* how to make an oak tree or how a human cell *knows* how to make us. Each cell carries its own DNA molecule within it with a unique sequence of bases, encoding the essential dogginess, oakiness, or humanness of us all.

In their historic paper's last line, Watson and Crick suggested that DNA's structure also provided a solution to the other great mystery of life. how biological information is passed on from one generation to the next – or why you look like your daddy. A major feature of the double helix is that wherever a base occurs on one strand of the DNA, a complementary base is found on the opposite strand: *A* is paired with *T* and *G* is paired with *C*. The pairs of complementary bases are held together by a hydrogen bond, a type of chemical bond. Chapter Five will look at chemical bonds in more detail, but a hydrogen bond is held together by the electromagnetic force existing between a positively charged proton on one base and the negatively charged electrons on its complementary base.

The information held in the DNA double helix is therefore redundant. The same information is held in two different forms: the coding strand (the strand that codes directly for proteins) and its complement. If one strand is removed, it can be used as a template to direct the synthesis of its complementary strand. This is exactly what happens when a cell replicates its DNA. The strands are pulled apart and each is used as a template to synthesize its complement. The enzymes involved in DNA replication examine the sequence of the single-stranded template and insert only the complementary base into the newly synthesized strand (in fact, fortunately for evolution, the copying isn't quite perfect – see below). After each strand has been copied, the pair of old and new strands form a duplex DNA molecule again. From a single parental DNA duplex, a pair of daughter duplexes is formed. One of the DNA duplex pair goes into one of the daughter cells and the other duplex goes into the other – biological information is copied. This simple mechanism underlies the replication of all living cells.

HOW DNA TELLS THE CELL WHAT TO DO

DNA encodes proteins but doesn't make them. That job is performed by structures inside cells called *ribosomes* which stitch together single amino acid units into strings which are called peptides if short, proteins if they are long. Left to their own devices, ribosomes might randomly string together amino acids, making totally random proteins. Ribosomes could make a staggering variety of proteins if allowed to function in this manner. Consider a relatively small protein, say only one hundred amino acids long. For each of the hundred positions in the protein there are twenty possible amino acids that could be inserted. There are therefore 20^{100} different ways of putting such a protein together. 20^{100} is an immense number. It means the product of $20 \times 20 \times 20 \times 20 \times 20 \times 20 \times \ldots 100$ times. For convenience, big numbers like this are usually expressed as a power of 10, so that they can easily be compared. In this system, 20^{100} can also be written as 10^{130}. For comparison, the number of electrons in the universe is a much smaller number, about 10^{80}, so there are not even enough electrons in the entire universe to count the number of possible 100 amino acid proteins! The ribosome's task is to make only a very tiny fraction of these possible proteins – the proteins the cell needs.

The problem is similar to house-building. The number of possible ways of putting several thousand bricks together is again a staggeringly large number and only a tiny fraction would amount to a functional house. The builder must select from the vast number of possible piles of bricks, one that corresponds to the desired house. He uses a plan that maps each brick (in principle if not in practice) to a specific position in space. The plan provides the builder with the *information* he needs to build the house. Similarly, the living cell must select from the vast number of all possible proteins the tiny fraction that corresponds to proteins with useful functions. The cell similarly needs to have some kind of plan or template and for this it uses DNA.

There is, however, (at least in animal and plant cells) a physical problem that must be overcome if DNA is to direct protein synthesis. DNA is held within the nucleus (a membrane-bound sac inside cells) of animal cells but the ribosomes are located outside it in the cytoplasm (the cellular material outside the nucleus). One possible solution would be for the DNA to pass through the nuclear membrane to the ribosomes where it is needed to direct protein synthesis. However DNA is a huge molecule, millions or even billions of bases long and it would not be easy for it to squeeze out through the membrane's small pores. What actually happens is that the information held in DNA is copied into a smaller, mobile analogue of DNA, known as RNA. RNA has all the same bases as DNA (well nearly all, it uses a base called uracil instead of thymine) on a sugar phosphate backbone, just like DNA. The only difference is that the sugar that goes into its backbone is ribose rather than deoxyribose (hence *RNA* rather than *DNA*). Since the bases are nearly the same, a single DNA strand can pair with a complementary RNA strand (the RNA uracil pairs with adenine) to form a DNA::RNA hybrid double helix. An enzyme called RNA polymerase then makes RNA copies of DNA genes. The RNA copy, called *messenger RNA* or simply *mRNA* is usually only one to several thousand bases long and can easily travel out through the pores in the nuclear membrane to reach a ribosome. It then locks into the machinery and tells the ribosome which protein to make.

The linear mRNA molecule is fed into a cleft within the ribosome which, acting like a barcode reader, reads the mRNA code, a codon at a time. The ribosome also has a docking station for amino acids. The ribosomes are, however, unable to identify or incorporate *naked* amino acids. Instead, each amino acid has to be tagged with yet another type of RNA molecule called *transfer RNA* or simply *tRNA*. Each tRNA molecule carries a three base anticodon sequence that acts as a comple-mentary barcode to identify the amino acid it carries. All the ribosome has to do is to pair up the two barcodes – the codon on mRNA with the anticodon on tRNA – and thereby insert the correct amino acid. The mRNA molecule is fed into the ribosome, a codon at a time, and at each step, the appropriate amino acid is clipped off the tRNA molecule and attached to the growing amino acid chain.

WHEN DNA DOESN'T WORK

The job of the DNA replication machinery is to make a perfect copy of the parental DNA strand. However, no copying is perfect – think how a photocopy of an image degrades as it is repeatedly copied in a photocopying machine. The DNA copying machinery of living cells is capable of a much higher degree of fidelity than a photocopying machine, but occasionally it does insert the wrong base. These errors are mutations that may or may not lead to changes in the amino acid sequence of proteins. Mostly the changes are of little or no consequence. Occasionally, if a mutation interferes with the function of an essential protein, it can be severely harmful, even lethal. The human genetic disease thalassemia is caused mainly by a mutation that changes a single amino acid in one of the globin proteins of blood haemoglobin. Even more rarely, mutations may provide some advantage to their host. Natural selection tends to favour organisms carrying advantageous mutations that allow them to produce more offspring. Over millions of years, organisms will evolve by selection of mutant offspring which are fitter than their parents. Mutations are therefore the elusive source of the variation that Darwin needed to complete his theory of evolution. They provide the raw material for all evolutionary change.

But where do these mutations come from? They are mostly generated during DNA replication. The enzyme that constructs new DNA strands is called *DNA polymerase*. Its basic activity is to string together units of nucleotides (the units of DNA that carry the bases) to make new DNA strands. However, the enzyme doesn't work unless it has a template on which to build the new strand. The old DNA strands are first unwound to allow DNA polymerase to slide along each strand (it forms a kind of doughnut structure around the DNA) to make the new strand. But occasionally (roughly one in every ten thousand bases) DNA polymerase makes a mistake and inserts the wrong base. The enzyme has several possible excuses for its errors. Radiation and some hazardous chemicals can cause mutations and do so by promoting errors during the DNA replication process. However, one source of replication error is unavoidable because it is due to the intrinsic *quantum nature* of the DNA code.

In *What is Life?* Erwin Schrödinger proposed that genes were aperiodic crystals (crystals lacking a periodic structure) and that quantum fluctuations may be a source of mutations. However, as he went along with the prevailing view that genes were made of protein, this suggestion was generally ignored when the true nature of the genetic code became known. The Swedish geneticist Per-Olov Löwdin from the University of Uppsala revived interest in the quantum nature of the genetic code by pointing out that it could be viewed as a linear array of protons. As described above, the coding properties of DNA are due to the hydrogen bonding between protons and electrons in the DNA bases: the position of these particles determines which hydrogen bonds can form and thereby the base-pairing underlying the genetic code. Protons and electrons are fundamental particles and their position is subject to quantum mechanics. The genetic code thus becomes a quantum code.

One of the peculiar features of quantum mechanics is that in most circumstances we cannot know exactly where a particle is: its position is uncertain (an aspect of Heisenberg's famous uncertainty principle that will be examined in more detail later). This uncertainty is the basis for a phenomenon called quantum tunnelling, whereby protons are said to tunnel from one position to another. In fact quantum particles don't really tunnel anywhere. It is just that the inevitable (Heisenberg) uncertainty in their position means that they materialize in places you wouldn't normally expect them.

The coding protons in DNA can (and must) quantum tunnel within the DNA molecule. This leads to *tautomeric* structures for DNA bases, with coding protons tunnelling from one atom to another to form modified chemical structures. Tautomeric forms of DNA bases can pair with the incorrect base: *A* can pair with *G* and *T* with *C* (rather than *A* with *T* and *C* with *G*). Watson and Crick proposed that if, during DNA replication, either the template DNA base or the incoming base is in the tautomeric form, then the wrong base may be inserted into the new strand, resulting in a mutation. Tautomeric forms of DNA bases account for about 0.01 per cent of all natural DNA bases, so incorporation of incorrect bases, due to tautomerization, is likely to be relatively common. However, our DNA replication machinery has *proofreading* enzymes able to recognize incorrectly inserted bases and clip them out of the growing strand. The inclusion of proof-reading into the system vastly reduces the error rate to only about one wrong base for every billion correct bases. Those errors that escape the correction machinery are the source of naturally occurring mutations; and their source is quantum-mechanical.

Watson and Crick's structure was therefore the culmination of centuries of biological progress. The great mysteries were laid bare: how biological information is encoded, how it is inherited and how it is changed. But it also pointed in quite a surprising direction, towards the involvement of that other great triumph of twentieth century science – quantum mechanics – in the fundamental basis of life and the driving force of evolution.

4

How Did We Get Here?

We all need to place our lives in some kind of historical context, to know where we come from. Ancestor worship is one of the most ancient forms of religion and the same craving finds a modern expression in the current popularity of genealogy. A relative of mine, John McFadden, recently traced our family back to one Cornelius McFadden. Cornelius owes his fame to being caught (probably around 1760) stealing a sheep on the island of Arranmore[1]. Sheep stealers were hanged in eighteenth-century Ireland but the authorities could however show some mercy, since the sentence could be commuted for men with families. Fortunately for Cornelius, his wife Nancy was heavily pregnant so my great-great-great grandfather suffered the lesser punishment of having his ears cut off. His wounds bound, he and his young wife were placed on a raft and pushed out to sea on the ebb tide, with only a single oar. The pair rowed along the coast, finally beaching on the island of Innishirther – a rather bleak and inhospitable place, but uninhabited. They settled there and thrived, raising eleven children. Their descendants migrated to the mainland, giving rise to a long line of McFaddens, including eventually me. The punishment suffered by Cornelius is remembered in a Gaelic phrase, 'Thug said Oidhe Concubhair air' (roughly translated, *the justice of Cornelius*) that commemorates their cruel punishment.

The tale of Nancy and Cornelius gives me some connection with the past but a few hundred years are very shallow roots in a history of life that stretches back billions of years. To find more – truly ancient – roots we must dig deeper into the past.

THE HOOF-PRINT OF EVOLUTION

The standard textbook illustration of evolution is the development of the modern horse. Darwin's friend and colleague Thomas Huxley worked out its sequence from more than two hundred species of fossil horses from Europe and America, using it to champion Darwin's theory. Horses are members of the family *perissodactyles* (animals with odd-numbered toes) that also includes rhinos and tapirs. The first perissodactyles were dog-like browsing animals weighing only about twenty kilos that first appeared in the North American forests roughly sixty million years ago. The subsequent evolution of the horse is thought to have been in response to a changing environment as woodlands gave way to open savannah. Gradually, over millions of years, many new species appeared with fewer toes, longer legs adapted for fast running and stronger jaws with big teeth adapted for grazing. Each new species was only slightly modified from its likely progenitor, but over millions of years there was a gradual increase in size and parallel changes in bone structure. Most of the new species became extinct, particularly during the last great Ice Age, but seven species of modern horse survived, which include the domesticated horse, asses and zebras.

The standard interpretation of the fossil record of horses, and indeed other animals, is one of gradual evolution. At any point in time there would have been many individual horses, all slightly different. Natural selection would have favoured the more successful variants so that, over the course of many thousands and millions of years, there would have been a gradual shift in horse shape, size and toe bones, to suit their new environment.

TO BE OR NOT TO BE, WITH HALF AN EYE

A favourite argument of anti-evolutionists is that complex structures, such as the mammalian eye, could not possibly have evolved by random mutations. To make this point, a metaphorical monkey is often recruited to bang away at a typewriter, typing in random characters. The question is then asked: how long would it take our simian typist to type out the whole of *Hamlet*? The answer can be fairly easily calculated. He would have to type out about 25^{30000} words (24 letters plus the space key raised to the power of the number of words in the text) to have a reasonable chance of typing in the correct text. If the monkey were a fairly proficient typist, say, one hundred words a minute, it would take him 25^{30000}, divided by one hundred, so approximately 10^{40000} minutes to hit the keyboard the requisite number of times. The number of minutes since the Big Bang are however a mere 10^{21}, a number vastly smaller. If we had a cosmic army of monkeys, one for every single electron in the universe and they had all been typing merrily away ever since the Big Bang, they would not have had sufficient time to achieve a tiny fraction of this feat.

However, the odds are radically transformed if we move from an entirely random selection of keys to adding one extra ingredient, selection. Imagine that we start with a small army of monkeys (a few hundred) and allow each to hit the keyboard once, selecting for breeding only those that correctly typed the first letter of *Hamlet*, W. Now, suppose that the ability to type that one letter is inherited, so that the progeny of these W monkeys invariably type the letter W with their first bang on the keyboard. Their next attempt at literary creativity – the next letter – would again be random, but once more we select for breeding only those that type H, the play's next letter. (Once again, saying that the ability to type the second character is inherited in the same manner as the first.) Continuing this breeding policy for just nine generations would breed monkeys that would competently type the first line: 'WHO'S THERE' (we will allow ourselves to add in the punctuation). If we continued with our breeding programme, allowing about ten years for each generation, then it would take a mere 300,000 years for us to breed a line of Shakespearean monkeys, able to type the entire text of *Hamlet*!

The odds are so much better because we have introduced selection into the random process. The only difference between this and Darwinian evolution is that the selection we have introduced is artificial – we are doing the selection. In nature, it is the environment that does the selection: natural selection. Perhaps regrettably, the ability to type lines of Shakespeare is unlikely to impress many female monkeys and would not cut much mustard in monkey society. A tropical forest environment is unlikely to favour a line of literary monkeys. The ability to see is, however, vital. A monkey equipped with sharper eyesight might be more successful at finding fruit with which to tempt prospective mates. It may more readily spot the attack of a rival. The monkey eye is thereby subject to Darwinian natural selection and it is this that shrinks the odds of developing the eye's complex structure from essentially zero to something achievable within geological lengths of time.

The key to the feasibility of this evolutionary scenario is the existence of a selective advantage for each and every step from simple to complex. This point is crucial to Darwinian evolution. The eye could only evolve if all the precedents to its modern form were viable and *each had an advantage over its predecessor*. Creationists claim that this is the weak point of the argument, for before an eye can reach the complex structure of its modern form, it must evolve through a thousand intermediate stages. But what use is half or a tenth of an eye? Surely an eye is only useful when all its parts are present and functioning?

This is a surprising claim, since Darwin himself used the eye to illustrate how the evolution of complex structures was indeed feasible by natural selection. In his essay, 'Organs of extreme complication and perfection', Darwin pointed out that far from half an eye or a tenth of an eye being of no value, there are many living animals with *half an eye* or a *tenth of an eye* which manage very nicely with their supposedly imperfect vision. Many microbes, including photosynthetic bacteria, possess the most rudimentary vision. These bacteria are able to swim towards bright light where their photosynthetic skills are most effectively deployed. They even have colour vision since they are able to concentrate where in the spectrum their chlorophyll absorbs the most light. Mutants can be isolated that lack this *phototactic* ability, demonstrating that the gene for some kind of photoreceptor is encoded in their DNA and thereby subject to mutation and natural selection. Whether bacterial vision represents a tenth or even one-hundredth of an eye is a matter of opinion but it certainly gives the bacteria a selective advantage over blind mutants. It is even possible that light sensitivity did not originally evolve with a role in vision at all. Many primitive organisms have light-sensitive proteins that are used to set their circadian (the biological rhythms that track night and day) clocks. It may be that this clock-setting function of primitive eyes arose well before their value for seeing was harnessed.

From its origins as a single photoreceptor protein in a bacterial cell wall, the next step towards the eye may be the patch of light-sensitive cells found on the body surface of some starfish, jellyfish, leeches and worms. These animals are unable to form an image but can respond to different levels of light and darkness, which may allow them to locate the brighter, and more productive, shallow waters and rock-pools. More complex light receptors are found in limpets, clams and flatworms in which the photosensitive cells form a shallow cup used to detect the direction of light. The obvious next step was to add some kind of focusing mechanism to form a simple image. Some molluscs achieve this by the pinhole camera principle – light is forced to travel through a narrow aperture that focuses the image onto a cup of light-sensitive cells, which we may now call the retina. Vertebrates, insects and octopuses instead incorporate a transparent lens that allows more light into the eye, yet focuses the image. Finally, a variable aperture pupil might be added to control the amount of light allowed in; thus we have the mammalian eye.

The key to this evolutionary scenario is its gradualism. There are however a group of eminent palaeontologists who challenge it. Stephen Jay Gould and Niles Eldridge point out that the fossil record does not actually record gradual changes in species. Instead most species, including most horses, appear abruptly in the fossil record, change very little over their entire history and then disappear just as unceremoniously. This pattern is well known to palaeontologists who have usually attributed it to the imperfection of the fossil record: the missing links between one species and another have all died without the decency to leave their remains as fossils. Yet recent exhaustive studies of well-preserved species, such as marine snails, tend to support the view that, generally, evolution seems to hop and jump, rather than crawl.

Gould and Eldridge claim that the punctuated pattern of change is a real phenomenon which reflects two rates of evolution. The first, *stasis*, is exemplified by *living fossils* like crocodiles that have changed very little or not at all for millions of years. The second pattern of evolution occurs more sporadically and is characterized by geologically instantaneous *speciation events* (sometimes called *macroevolution*) in which one or several new species are generated. The pattern of long periods of stasis interspersed with rapid spurts of evolutionary innovation, they term 'punctuated equilibrium'. Evolution that goes at two different speeds clearly needs some kind of gearing mechanism to change from one to another. Gould and Eldridge suggest that when evolution makes a jump, natural selection may be acting at a higher level to select whole species or groups of species, rather than individuals. However, many other evolutionary biologists, such as Richard Dawkins, take great exception to the view that natural selection acts on any unit higher than that of an individual (or even a single gene).

Whether evolution proceeds by tiny steps or big leaps, by examining the fossil record we can trace the evolution of man and animals back to the emergence of the first animals in the *Cambrian explosion* five hundred and fifty million years ago. Rocks earlier than the Cambrian explosion have very few fossils – which are nearly all microbes. Unfortunately, microbial fossils are not very distinctive. They give few clues about the evolutionary changes that led to the emergence of animals. To go deeper into the history of life we need to dig into DNA, rather than rocks.

THE GENE CLOCK

The word for milk is *lait* in French, *latte* in Italian, *leche* in Spanish and *leite* in Portuguese but *milk* in English *milch* in German and *mjölk* in Swedish. French, Italian, Spanish and Portuguese are all Italic languages; whereas English, German and Swedish are Germanic languages. Other European language groups include Celtic, Hellenic and Slavic. In 1786, Sir William Jones, an English judge serving in India, first noticed similarities between the ancient Indian language, Sanskrit and various European languages. For instance, the word for king is *rex* in Latin, *ri* in Irish, *raja* in Sanskrit. The same root turns up in the English word ruler. Sir William considered that these similarities could not have arisen by chance but must reflect a common linguistic inheritance. The English scholar Thomas Young later coined the term Indo-European to describe these common languages.

Modern languages are thought to be derived from an ancestral *proto-Indo-European* language spoken by either a Bronze or Neolithic Age people. The original Indo-Europeans would have spoken a common proto-Indo-European but gradually as the people dispersed, their languages diverged to develop into the modern family of languages. Philologists (those who study language development) compare similar words in each language to derive a plausible ancestral word. For instance, a single word for milk, approximating to *lakte*, is thought to have been used by people who spoke proto-Italic, the ancestral language of modern French, Italian, Spanish and Portuguese. The patterns of divergence in each language group could then be estimated by counting the number of sound shifts required to change from the putative ancestral word to all its modern forms. Languages linked by few sound shifts, such as Spanish and Portuguese, are considered to have diverged relatively recently. Languages linked through more sound shifts, such as German and Portuguese, must have separated much further back. In this way, a family tree of languages can be suggested. By dating language divergence to a historical event (for instance, the settling of England by an Anglo-Saxon speaking people in about 550 AD that led to the separate development of English), philologists can provide a very rough calibration of the rate of divergence of languages.

People's common inheritance can also be traced in their genes. Consider a short gene segment from four different individuals that reads: ATTGC in Harry, AATCA in Jim, GATGC in Betty and ACTGC in Bertha. A plausible sequence that might have belonged to their last common ancestor (the proto-Harry-Jim-Betty-Bertha) would be *AATGC*, since only one base change is needed to change the ancestor sequence into any of its modern descendants. Harry, Jim, Betty and Bertha could be said to belong to a gene family. If the DNA of another individual, Ted, was sequenced as AATTT we then could conclude that Ted was more distantly related to Harry, Jim, Betty and Bertha since two sequences changes are required to connect his sequence to any of the others. The last common ancestor of all five individuals, the proto-Harry-Jim-Betty-Bertha-Ted must have lived earlier than proto-Harry-Jim-Betty-Bertha. Just as with words, DNA sequences can be used to draw up family trees, only this time the tree reflects genetic rather than cultural inheritance.

Carl Woese of the University of Illinois was the first to make extensive use of DNA sequences to examine the early evolution of living creatures.

He used the gene sequences encoding one of the sub-units of the ribosomes (the protein-making machine in cells) as a gene clock to construct a universal genetic tree. The tree divides all life into three domains. The first contains the eukaryotes (whose DNA is enclosed within the nucleus) which includes the unicellular protozoa (such as amoeba) and multicellular plants, fungi and animals – and us. The other two domains both consist entirely of prokaryotic (which means *before* the nucleus – whose DNA is not enclosed within nuclei) organisms. The eubacterial (*true* bacteria) domain contains most of the bacteria we are familiar with, such as *E. coli*. The third domain is that of a newly recognized bacterial group, called the *Archaea*.

Many aspects of the tree agreed more or less with evolutionary thinking. The ribosomal RNA sequences of the multicellular animals' groups (vertebrates, worms, sponges, arthropods etc.) diverged at roughly the time of the Cambrian explosion. The rRNA of plant chloroplasts (which have their own DNA including rRNA genes) was found to be similar to bacterial rRNA, tying in with a theory championed by Lynn Margolis in the late 1960s, that these organelles were descended from symbiotic bacteria. The separate deep branching of eukaryotic (animals and plants) genes did come as something of a surprise. Until then it was generally assumed that eukaryotes had branched off from some bacterial ancestor billions of years ago; but that would have left us closely related to one of the bacterial groups. There was no evidence for this in the tree. Eukaryotes appeared as ancient as bacteria. This feature of the universal tree still remains a puzzle.

The recognition of the Archaea as a distinct domain of life also came as a major surprise to biologists. We have already met some of the Archaea in Chapter Two. The extreme thermophilic bacteria thriving in the undersea vents; the halophiles living in briny waters; and the methane-producing bacteria, are all Archaea. They have markedly different enzymes, fats, and cell structure to eubacteria and eukaryotes. Scientists had until then considered them as bacterial oddities but Woese's analysis placed them as a separate form of life that, at a molecular level, is quite as different from, say, *E. coli* as we are. Scientists have sequenced all 1,664,976 DNA bases that make up the genome of an Archaea called *Methanococcus janaschii*, fished out of a two thousand, six hundred metre deep 'white smoker' hydrothermal vent chimney. Many of its genes are similar to those found in eukaryotes, suggesting that our nuclear genome may be the descendant of an ancient archaeon.

HOW DID GENES EVOLVE?

By and large, gene sequence data supports the neoDarwinian notion that gene evolution has involved a series of gradual modifications of existing genes through mutations. Nevertheless, problem areas remain. The first (already mentioned): how to account for apparent big jumps. A related problem, apparent in the DNA record, is the relationship between the major protein families. Examination of genes from diverse organisms has established that all modern proteins fall into about a thousand distinct protein families'. Although evolution *within* protein families, such as the globin (the protein in haemoglobin) gene family, can generally be traced through a number of antecedent proteins present in living creatures, finding the links *between* the protein families is much more difficult. Animal globins bear some relation to oxygen storage proteins found in bacteria but there is little or no identifiable relationship between these globin-related proteins and any of the nine hundred and ninety-nine or so, other protein families. The same is true for all the other protein families – there is much evidence for Darwinian evolution within the family, but no obvious close relative from which the family could have evolved. Each protein family is like a separate galaxy (of related proteins) in a vast outer space of protein sequences. New protein families must have arisen from existing proteins by some kind of mutational process but how their sequence traversed this vast empty sequence space devoid of Darwinian intermediates, is a mystery. It seems molecular evolution often proceeds though a series of small steps but that sometimes it takes big leaps – rather like the punctuated evolution envisaged by Gould and Eldridge. Big leaps are big problems for neoDarwinian evolution because the chances of a big jump landing anywhere useful are generally thought exceedingly small. As Richard Dawkins states, 'However many ways there are of being alive, it is certain that there are vastly more ways of being dead . . .'[3]

Another problem for the neoDarwinian process is the evolution of metabolic pathways. This is a kind of molecular version of the eye problem – how to evolve complex structures – but its solution is not as apparent as the evolutionary pathway that led to the eye. The basic problem is that the complexity of biochemical pathways (unlike the eye) do not appear reducible. For instance, one of the cell's essential biochemicals is AMP (adenosine monophosphate) the precursor of ATP (the energy-carrying molecule) which also finds its way into DNA, RNA and many other cellular components. AMP is made from ribose-5-phosphate, but the transformation involves thirteen independent steps involving twelve different enzymes (which we will represent as: $A \rightarrow B \rightarrow C \rightarrow D- \rightarrow E \rightarrow F \rightarrow G \rightarrow H \rightarrow I \rightarrow J \rightarrow K \rightarrow L \rightarrow M$ where A is ribose 5-phosphate and M is AMP). Each of the twelve enzymes involved in this pathway is absolutely essential for the biosynthesis of AMP. Darwinian evolution would require this complex system to have evolved from something simpler. But, unlike the eye, we cannot find the relics of simpler systems in any living creatures. As far as we know, nothing simpler works. Half or a quarter or a twelfth of the pathway does not generate any AMP or indeed anything else of value to the cell. It appears that the entire sequence of enzymes is needed to make any AMP. But without viable stepping stones, how can the entire complex system have evolved through Darwinian natural selection?

One explanation often cited is that complex biochemical pathways have evolved backwards. The story goes that the primitive cell initially utilized the final biochemical in the pathway (M or AMP) directly, as it was already available in the primordial soup. However, as primitive cells used up the supplies of M, any cell that evolved the capability of making M from another available biochemical would have had a selective advantage. One of those biochemicals was L, and a cellular innovator soon evolved an enzyme which could perform the transformation of $L \rightarrow M$. Eventually supplies of L were, in their turn, depleted, creating selective pressure for a second evolutionary step to acquire the enzymatic capacity to make more L from one more readily available biochemical, K. Eventually, the entire pathway: $A \rightarrow B \rightarrow C \rightarrow D \rightarrow E \rightarrow F \rightarrow G \rightarrow H \rightarrow I \rightarrow J \rightarrow K \rightarrow L \rightarrow M$, evolved through this series of backward steps.

The problem with this explanation is that it requires all of the intermediate biochemicals (B,C,D,E,F,G,H,I,J,K,L) to have been sloshing about in the environment of the primitive cell. Yet AMP is a ribose sugar, that goes into making RNA. We will soon be exploring the enormous difficulties in making ribose sugars by the kind of inorganic chemical processes going on within the early Earth. As you will see, it is unlikely that even one, never mind all eleven, of the biochemicals in the pathway from ribose 5-phosphate to AMP was present in any significant quantity in the environment of the primitive cell.

In his book *Darwin's Black Box*, the biochemist Michael J Bethe of Lehigh University considered the evolution of the AMP pathway and several other complex biochemical systems such as the mechanism of the bacterial flagella. Bethe contends that these complex biochemical systems cannot be broken down into a series of (forward or backward) steps subject to Darwinian evolution. He maintains that their evolution is totally inexplicable in terms of standard evolutionary theory or indeed any scientific theory. Bethe's solution is either radical or hopelessly archaic depending on your point of view; he proposes that the existence of complex biochemical systems is due to God's intervention in the evolution of life on Earth.

A third problem for the standard evolutionary theory is the apparent existence of a *heretical* type of mutation. Standard neoDarwinian evolutionary theory predicts that mutations occur randomly with no respect to the direction of evolutionary change. Natural selection provides the direction of evolution by selecting hosts with beneficial mutations; but those mutations are generated randomly. This does not mean that all sites in DNA have the same mutation rate. In fact we know that there are regions in most chromosomes that are highly mutable (probably due to their local chemical environment, or because they bind mutation-promoting proteins). However, it does mean that the mechanisms that introduce mutations into DNA are presumed not to *know* which bases are likely to generate advantageous mutations if they mutate.

Nonetheless, when John Cairns of the Harvard School of Public Health in Boston set out to test this reasonable premise he found things were not quite so simple. Cairns incubated cells of the common gut bacterium *E. coli* in conditions where a single mutation could rescue them from starvation. He used an *E. coli lac⁻* strain deficient in an enzyme called β-galactosidase (β-gal) needed for the cells to eat lactose (milk sugar). He then fed the cells on a diet of only lactose. Without β-gal he expected all the cells to starve. In fact it takes a lot to kill *E. coli* by starvation, mostly the cells just shut up shop and go into what is called a stationary phase, where either they don't replicate or only very slowly. *E. coli* cells can survive for many weeks in this stationary phase.

Cairns fed a parallel culture of *E. coli* cells on yeast extract, which the cells could eat without need for the β-gal enzyme. The standard neoDarwinian theory would predict that the mutation *lac⁻* → *lac⁺*, to generate a fully functional β-gal enzyme, should occur at the same rate for the cells fed on yeast extract, compared with cells on the starvation diet of lactose. The only difference should be that, for the cells fed only on lactose, the mutation would rescue them from starvation; whereas the mutation would be irrelevant for the cells happily feeding on the yeast extract. What Cairns actually found was a much higher rate of *lac⁻* → *lac⁺* mutation when the cells had only lactose to eat. Cairns examined other genes but their rate of mutation was unchanged by starvation, indicating that the phenomenon was not caused by a general increase in mutation rate.

These adaptive mutations suggested that a starving cell could sense that it was starving and somehow *choose* the gene it needed to mutate to save itself from starvation. Cairns' paper describing adaptive mutations was published in *Nature* in 1988[4], unleashing a storm of controversy. The difficulty was that there was no known mechanism which could allow the environment of a living cell to influence the targeting of DNA mutations. The direction of information flow in the cell is from DNA through RNA to protein and outwards to the environment. There is currently no known mechanisms by which information can flow backwards from the environment to DNA to account for these mutations.

Since 1988, hundreds of publications have appeared that have either supported or denied the phenomenon of adaptive mutations. Adaptive mutations have been proposed to occur in many types of bacteria as well as more complex yeast and animal cells and have even been implicated in cancer. One of the most impressive demonstrations of the phenomenon was by Barry Hall of Rochester University who demonstrated that two sites just a few bases apart on the same DNA molecule could be subject to widely different mutation rates, dependent on whether or not those mutations were adaptive[5]. Whatever their mechanisms, adaptive mutations appear to be able to bias the mutational process to favour certain genetic changes.

*   *   *

I must emphasize that though there may be some doubt concerning the mechanisms involved in evolutionary change (and there are likely to be many), this should not be confused with any doubt concerning the process of evolution itself. There is overwhelming molecular evidence that all modern species have evolved from earlier species. Indeed, there is considerable evidence that we have all evolved from a single ancestral cell. Let us next examine the probable nature of that common ancestor.

THE PROTO-CELL

Molecular evolutionary studies have established the existence of the major domains, but what kind of creature was the ancestor of the Archaea, eubacteria and eukaryotes: the proto-cell? That there was indeed a single common ancestor of all cellular life is indicated by the number of features we all have in common (DNA, RNA, ribosomes, the genetic code); but what did that proto-cell look like? How did it live?

We can make an estimate of how many genes the proto-cell is likely to have possessed by counting the number of genes found in all three domains of life. The simplest explanation for the presence of common genes is that they reflect a common inheritance from the proto-cell. Scientists who have examined genes from each of the three domains estimate that there are eight hundred to a thousand *ancient conserved regions* (ACRs) in modern proteins.[6]

Surprisingly perhaps, eight hundred to a thousand genes is a little more than the number of genes present in the genome of a living bacterium, *Mycoplasma genitalium*. This organism causes non-gonoccocal urethritis (inflammation of the urethra which is not gonorrhoea) and respiratory infections in humans, and has the smallest known genome of any living creature. Its entire chromosome of 580,070 DNA bases has been sequenced and found to code for only four hundred and seventy proteins. However, this microbe is not really a free-living organism – it lacks the enzymes necessary to make many essential biochemicals. It barely manages to replicate under very cosseted laboratory conditions. It is actually a highly evolved parasite that relies on our cells to do much of its biosynthetic hard work. It is not a feasible proto-cell. Nevertheless we will accept a lower estimate of about five hundred genes as the minimum number of genes likely to have been present in the last common ancestor of all cellular life, the proto-cell.

But what was the proto-cell like? We can approach this question by examining the characteristics of the deepest branches of the rRNA tree. For instance, the deepest branching Archaea are thermophilic bacteria that *breathe* sulfur compounds. The deepest branches of the eubacterial domain are thermophilic (heat-loving), sulfur-utilizing photosynthetic bacteria; suggesting that the *proto-eubacterium* was an anaerobic photosynthetic cell. Most researchers do not think it likely that photosynthesis evolved very early in life. It is a highly complex reaction that requires the co-operative action of many enzymes. The more ancestral characteristics are likely to be closer to the Archaea that use the energy of sulfur compounds directly to fix carbon dioxide. On this basis, the likely characteristics of the proto-cell were probably similar to the Archaea, suggesting it lived in a hot, sulfurous, oxygen-free environment and it used sulfur to fix carbon dioxide.

But when did the proto-cell live? Do the conditions we have described correspond to the likely environment where it emerged?

THE FIRST FOSSILS

The first undisputed evidence of life is in rocks about three and a half billion years old. Large circular structures called stromatolites (from the Greek meaning 'stony carpet') are present in fine-grained flint-like sedimentary rock[7] called chert. Some of these rocks contain curious white concentric rings about a metre across. Nobody knew the origin of these structures until in the 1950s scientists cut sections of the rock and examined them under the microscope. They found rods, spheres and chains of cells, unmistakably the fossilized remains of microbes. The oldest stromatolite microfossils were found by the University of California's paleobiologist J. William Schopf, in chert outcrops near Chinaman Creek in Western Australia dated to three and a half billion years ago. Schopf found many short rods and filaments in the rock very similar in size and shape to the cells of modern bacteria, including the photosynthetic

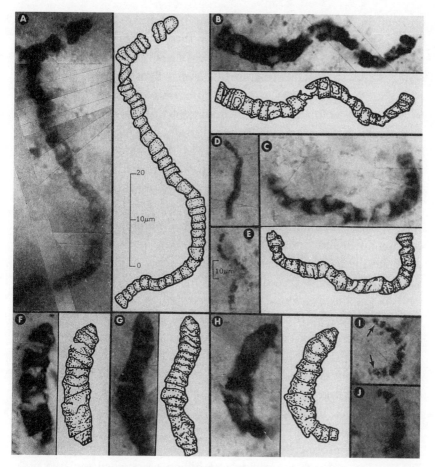

FIG 4.2   Terrestrial microbial fossils. Carbonaceous fossil microorganisms
(with interpretive drawings) of the Early Archaean (about 3,500 million years old)
Apex chert of northwestern Australia. From J. William Schopf 'Microfossils of the
Early Archaean Apex Chert: New Evidence of the Antiquity of Life' (1993)
Science 260: 640–646.

cyanobacteria of Chapter Two. Some of the fossils were incredibly well
preserved, allowing Schopf to discern their internal structure and even
recognize cells that appeared to have been undergoing cell division before
their lineage was abruptly petrified. Cells of various sizes and shapes
were present suggesting that a diverse microbial community was already
thriving.[8]

Rocks even more ancient than those of Chinaman Creek are found on Isua in West Greenland. These date back about 3.85 billion years. Nobody has yet found any microfossils in these rocks but this is unsurprising since they have gone through episodes of heating at high pressures that would have destroyed the structure of any ancient organisms. Another way of looking for evidence of past life is to examine the ratio of carbon-12 to carbon-13 isotopes[9] in sedimentary rock. Plants and bacteria that convert inorganic carbon (as carbon dioxide, carbonate or methane) to biomass discriminate against carbon-13 so their tissues are enriched by the lighter carbon-12 isotope. Rocks that hold the fossilized remains of ancient carbon-fixing plants and microbes preserve this preponderance. Microfossil-bearing rocks including the ancient cherts show the unmistakable signature of biological carbon fixation. Although the Isua rocks contain no fossils they do have grains of graphite within them which could be the roasted remains of ancient microbes. Scientists from the Scripps Institution of Oceanography, University of California, examined the Isua graphite and found enriched levels of carbon-12, indicating that the carbon may possibly be of biological origin. The levels were not as high as in later rocks, possibly due to high temperature re-equilibration of the isotope ratios, but do suggest a biological origin.

The present consensus is that by three and a half billion years ago, diverse assemblages of bacteria, not visibly different from modern forms, were swimming in the primordial seas. A community of bacteria must surely have taken many millions of years to evolve from simpler beginnings, so the proto-cell, the ancestor of the diverse forms, must have lived earlier, perhaps as early as 3.85 billion years when the Isua rocks were being laid down.

THE BEGINNING

Modern cosmologists tell us that the universe came into existence some thirteen to fifteen billion years ago in the *Big Bang*. Moments later, the universe was at a colossal temperature of tens of billions of degrees but squeezed into a space much smaller than an atom. A period of tremendous expansion, lasting less than a billionth of a second, followed, to generate a universe, roughly the size of a football. Temperatures were still too hot for atoms to exist but the football universe continued to expand at a more leisurely pace cooling slowly until, after about a hundred thousand years, conditions were cool enough for protons, neutrons and electrons to combine to form atoms, mostly of hydrogen and helium. Over the course of millions of years, gravitational forces pulled those atoms closer together so that they condensed to form stars and galaxies. The primordial stars continued to collapse, increasing their internal temperature until they were so hot their atomic nuclei fused. Nuclear fusion released enormous quantities of energy as radiation, and the stars burst into light.

It is not known whether the first round of star formation generated any planets, but if it did, they could not have harboured life. Nearly all matter was in the form of just hydrogen and helium so only a very limited chemistry – certainly not enough to generate life forms – would have been possible. Heavier nuclei like carbon, nitrogen and oxygen were being made by nuclear fusion reactions inside stars, but these were only released into cooler spaces when a star exploded in a supernova. The dust and gas from the first supernovae went through another round of condensation and star formation before exploding again to release even heavier elements such as iron, sulfur and phosphorus into the universe. These condensed once again to form both stars and planets. Our own sun is one of these third generation stars. It seems a bright beginning for life to know that the stuff of all of us was made in stars and scattered throughout space by supernovae. We are, quite literally, stardust.

About four and a half billion years ago, a particular cloud of gas and dust left over from an earlier supernova condensed to form the sun. Remnants of the condensate that were scattered further afield collected in rocky agglomerates that formed the planets. It is generally believed that Earth's present crust was formed about four billion years ago, probably by an up-welling of molten magma from it's core. The Earth's atmosphere at this time was mostly hydrogen, but this was blown away by an intense period of solar wind, leaving a dry airless planet unsuitable for life's emergence. Fortunately for life, the crust was pockmarked with active volcanoes belching out huge quantities of gas and water vapour to form the primordial atmosphere. What this consisted of is a critical question for origin-of-life theories as it determines the kind of prebiotic chemistry possible on the Earth's surface. Unfortunately, we don't know its composition. The current best guess is that it would have been similar to the gases released from modern volcanoes, with carbon dioxide, carbon monoxide, nitrogen, hydrogen chloride and water vapour predominating, and only trace amounts of oxygen and hydrogen.

For several hundred million years, leftover rock fragments showered the Earth from space in what is called the intense late bombardment. The scars of this can be clearly seen on the craters of the moon but on Earth they have been erased by tectonic activity and erosion. This bombardment was so powerful that the heat generated by its massive impacts would have periodically vaporized all surface water and extinguished any primordial life. It wasn't until 3.8–4 billion years ago that most of these roving rocks were swept out of nearby space and the Earth cooled sufficiently to allow water to rain down and fill its oceans.

But, as described above, there is evidence for biological carbon-fixing life in rocks 3.85 billion years ago: this appears to place the origin of life back at the tail end of the late bombardment. If the evidence from the Isua rocks is to be believed, then, as closely as the geological record can tell us, life emerged on Earth as soon as it was possible. The rapid appearance of life on Earth places an important constraint on theories to account for its emergence. We cannot, I believe, rely on extremely unlikely scenarios to get us out of our difficulties. Life's rapid emergence implies that once conditions were suitable, life was probable.

Now we come to the crux of the problem. Where did the proto-cell come from and how? The standard scenario (which we will examine more closely below) envisages life spontaneously emerging from a primordial soup of chemicals in the ancient ocean. Somehow, we must account for the spontaneous emergence of a cell with enzymes, ribosomes, RNA and DNA with all the 500 or so genes of the proto-cell. How did this structure emerge from the chemical soup?

I am sure there's no need to remind you of all those busily typing monkeys to persuade you that an organism with five hundred genes, each made up of about a thousand DNA bases could not have arisen entirely by chance. A billion universes each populated by billions of typing monkeys could not type out a single gene of this genome. Hoyle and Wickramasinghe (of whom more later) describe the likelihood of the event as equivalent to the chances that a tornado sweeping through a junkyard might assemble a Boeing 747.

The simplest living cell could not have arisen by chance. Just like the eye, the proto-cell must have evolved from simpler ancestral cells, presumably by a process of natural selection. But this is where the first big problem with the origin of life arises. What were those simpler entities? Darwinian evolution depends on its gradualism. Each small step in the evolutionary ladder must be viable and each must represent a tiny improvement on its progenitor. This is why we do not need to examine the fossil record to find the antecedents of the modern eye. They are all around us. Those simpler eyes are still in use precisely because they are viable structures that do the same job of seeing today as when they first evolved millions of years ago. If the proto-cell arose by Darwinian evolution from simpler ancestors, each ancestor must similarly have been viable and each must have represented a small advance *selected* by the process of natural selection. What has happened to these ancestors of the proto-cell? Why don't we see any of them today? If microbes that lived 3.85 billion years ago could replicate with less than five hundred or so genes, why don't today's microbes survive with fewer genes[5]?

The problem in a nutshell is that today's microbes need at least several hundred genes to grow and replicate. We have no reason to believe that the proto-cell, the last common ancestor of all cellular life, used fewer genes. How did life make the leap from the primordial chemical soup to the proto-cell?

LIFE IN THE SOUP

The standard textbook explanation of the origin of life on Earth was independently put forward by the Russian scientist Alexander I Oparin and the English geneticist J. B. S. Haldane at the beginning of the twentieth century. It is often known as the Oparin-Haldane hypothesis. In the beginning, so it goes, the Earth's atmosphere was rich in hydrogen, methane and water vapour. When these components were exposed to various sources of energy, such as lightning, solar radiation and volcanic heat, the gases combined to form a mixture of simple organic compounds. Over the course of millennia, these compounds would have accumulated in the ocean to form a warm, dilute primordial soup that eventually yielded a new kind of molecule, one that could replicate itself, the *primordial replicator*. Still a very simple entity, the replicator would have generated many errors in its replication: mutations. The mutant replicators competed for materials with which to build more replicators. Those mutant replicators which were most successful left the most descendants, and a molecular Darwinian natural selection took hold to drive the replicators towards greater efficiency and greater complexity. Some replicators captured other molecules to aid replication: amino acids, peptides and metabolites to provide energy. Membranes were recruited to protect the primordial replicators from the vagaries of the external environment and eventually the first cell was born.

The theory's value was that it provided some kind of scientific framework to understand how life could have originated. It may not have been entirely plausible but, at least, it maintained the possibility of testing the hypothesis in the laboratory. However, the theory remained just that, a theory, for some decades until in 1952 a young research student, Stanley Miller, working in the Chicago University laboratory of Professor Harold Urey, performed experiments designed to test the theory. Miller simulated the atmosphere of the early Earth by filling a bottle with gases thought to have been present in the primordial atmosphere: methane, hydrogen, ammonia and water vapour. He simulated the ocean by filling the base of the bottle with hot water. Lastly, he simulated lighting by igniting the mixture with electric sparks. To Miller's surprise – and the astonishment of the scientific world – when he analysed the composition of his flask-ocean after a week of sparking his primordial atmosphere, he found that it contained significant quantities of amino acids, the building blocks of protein.

Miller's experiments were hailed as the first step in the creation of life in the laboratory and remain a landmark today. Although no self-replicating molecules were generated in his experiment, it was generally believed that given sufficient time and a sufficiently large ocean, Miller's primordial soup amino acids would have polymerized to form peptides and complex proteins, eventually yielding the Oparin-Haldane replicators.

Yet, more than forty years later, this hasn't happened. Scientists have not even come close to generating anything like a self-replicating molecule from any primordial soup experiment. To understand why, we need to look more closely at Miller's experiments. The first problem is the complexity of the chemical mixture that Miller generated. In most experiments, much of the organic material produced was in the form of a complex *tar*. Organic chemists are very familiar with these tarry mixtures. In my undergraduate days as a biochemist, we were coaxed through the synthesis of a few organic chemicals. A successful synthesis might generate a white powder or clear coloured liquid but, more often than not, my efforts were rewarded with the same kind of organic tar Miller produced in his bottles. The weary demonstrators each had their own word for this product of my chemical ineptitude: gunk, crud, muck, scum. None was in the least interested in its possible role as potential primordial soup. The problem with such mixtures is that it is notoriously difficult to get anything more than other gunk from them. Chemically they are not productive. They are so complex that any specific molecule, like a specific amino acid, tends to react with so many different compounds that it gets lost in a forest of inconsequential chemical reactions. Hundreds of thousands of undergraduates have been producing gunk for decades and nothing more interesting than gunk has emerged. What this kind of synthesis lacks in comparison to life is not complexity but *direction*.

The second limitation to Miller's experiments was that he obtained no polymers of amino acids – no peptides or proteins. The problem here is the very stuff of life itself, water. All biological polymerization reactions remove a water molecule from the reactants. This is very hard to do in a solution of billions of water molecules. It is the chemical equivalent of trying to squeeze one more commuter onto a packed train. The tendency is for the commuter or the water molecule to pop back out again. Biological polymers tend to spontaneously *hydrolyse* (a water molecule comes out of solution to attach to the polymeric bond and break it) and fall to pieces in water – though slowly. In living cells, enzymes polymerize amino acids by coupling polymerization to energy-yielding hydrolysis reactions. But enzymes are made by cells; we can't invoke enzymes to make the first cells.

Another problem is to do with the structure of many organic compounds. The molecule of a simple amino acid like alanine (one of the amino acids in Miller's soup) has two *chiral* (left- and right-handed) forms: L and D. As Louis Pasteur discovered in 1848, most biochemicals exist similarly in left- and right-handed forms, whose molecules form mirror images of each other (rather like a left- and right-handed glove). Yet the amino acids in Miller's soup were generated in equal amounts of L and D forms. The polymerization steps that could yield peptides and amino acids simply do not work when both chiral forms are present.

A final problem is that the ingredients he used to produce his primordial atmosphere were probably not primordial. Miller filled his bottles with a *reducing* atmosphere rich in hydrogen and the hydrogen-containing gases, methane and ammonia. One of the reasons he chose this kind of atmosphere was that he knew that if he utilized the opposite kind of atmosphere, an oxidizing atmosphere rich in oxygen, this would have prevented organic synthesis. Oxidization tends to break up organic molecules rather than form them. Although a reducing atmosphere was seen as plausible in the 1950s when Miller performed his experiments, it is now thought unlikely. Life probably emerged in an oxidatively neutral atmosphere, consisting mostly of carbon dioxide and nitrogen. When primordial soup experiments utilize these more plausible atmospheres, yields of amino acids and other organic compounds are severely reduced.

Since Miller's experiment, other experimenters have used many recipes to cook a great variety of different flavours of primordial soup. Each has synthesized one or more of the essential ingredients of life but usually with low yield and with most of these limitations. Nothing remotely approaching a self-replicating molecule has yet been produced.

LIFE FROM SPACE

Most scientists quite happily entertain at least the possibility of extraterrestrial life (as outlined in Chapter Two) but draw the line at alien abductions or flying saucers. However the concept that material from outer space made vital contributions to the development of life on Earth has been gaining increasing respectability in recent years. The British astronomer Sir Fred Hoyle (who coined the term *Big Bang*) and his colleague Chandra Wickramasinghe suggested that many of the problems associated with primordial soup synthesis on Earth would be cured in space. Amino acids, sugars and even nucleotides were proposed to have been abundant on the comets or meteorites that showered the Earth four billion years ago. These ideas received some support after amino acids and other organic compounds were detected in a meteorite that fell in

Murchison, Australia in 1969. When *Giotto* rendezvoused with Halley's Comet on 13 March, 1986, it discovered that it was made up of about twenty-five per cent organic carbon and about forty per cent of its mass was water. If comets are rich in organic compounds, then they could have delivered massive quantities of organic material to the early Earth that filled the oceans with the primordial soup.

The problem with this soup-from-space scenario is that simulations of cometary impacts suggest that all organic material would be burned up in the heat of the resulting fireball. An alternative delivery route is soft landings by interplanetary dust particles. Many thousands of tons of these tiny particles rain down on Earth each year and they contain about ten per cent organic carbon. How much of this organic material would survive the inevitable heating as these particles pass through our atmosphere is uncertain.

The most radical life-from-space theory, *Panspermia*, was first proposed by the Nobel prize-winning Swedish chemist Svante Arrhenius in the early twentieth century. Arrhenius suggested that life-bearing planets throughout the galaxy eject the spores of bacteria and other microbes into space. The spores were proposed to survive interstellar travel to alight upon a planet, germinate and thereby spread life. The idea did not attract much support but a new twist was added in the 1970s when Leslie Orgel and Francis Crick (of DNA fame) proposed the theory of directed panspermia, in which an alien intelligence is proposed to have travelled throughout the galaxy, sowing the seeds of life on suitable planets. Although the theory stimulated science fiction, it gained little support in the science community.

Hoyle and Wickramasinghe made a further radical proposal in the 1980s. They claimed that interstellar dust grains (tiny particles that congregate in regions of space to form interstellar clouds that eventually coalesce to form stars – stardust) are actually freeze-dried bacterial cells formed by some kind of interstellar evolutionary process. They based their extraordinary claim on the analysis of light from these grains; but hardly any other scientist agrees. Hoyle and Wickramasinghe went on to make ever more extraordinary claims: that key evolutionary innovations were initiated when new genetic instructions arrived from space; that cancer is caused by our cells picking up the wrong instruction from space; and the bacteria and viruses that cause infectious diseases arrive from space. Needless to say, their later theories have attracted even fewer supporters.

To a certain extent the wacky end of the life-from-space field has detracted from its more plausible possibilities. Simple organic compounds are present in space and (it seems to me) that their relative abundance is likely to have played a role in delivering some of the ingredients of the primordial soup. But this still leaves a big gap from soup to cells. The far more speculative panspermia or directed panspermia ideas aim to bridge this gap, yet all they really achieve is to shift the problem. If life came from space then it must have evolved elsewhere. But why should any other planet be a more plausible environment for the origin of life than Earth? A very useful principle was first articulated by a fourteenth-century monk, William of Occam. The principle, known as Occam's razor, states that you should not opt for a complex solution to a problem if a simpler exists. Applying Occam's razor to the problem of the presence of life on Earth, the good monk would surely instruct us to look for life's origins on the same planet on which we find life.

FEET OF CLAY

Whereas most origin-of-life scenarios unfold in the primordial sea, Graham Cairns-Smith of Glasgow University moved his origin onto the land, or more precisely, to clay. Clay is a very complex material composed of finely ground grains of minerals and mineral crystals. The structure and properties of crystals have always tantalized scientists interested in the origin of life, because of their ability to mimic many properties of living organisms. If you drop a salt crystal into a salt solution, the crystal will grow by depositing the salt onto the sides of the crystal. If the solution is already saturated with salt, then the crystal will replicate itself as thousands of similar crystals and instantaneously drop out of the solution. Cairns-Smith suggested that life started off with mineral crystals replicating within clay.

Cairns-Smith's theory certainly solves the problem of the availability of prebiotic materials, as clay was likely to have been hugely abundant on the early Earth. He proposed that the charged surfaces of clay crystals played the role of genes to allow the clay replicators to modify their environment and enhance their replication. Defects in crystal structure would have been inherited by daughter crystals, allowing a Darwinian evolutionary process to select for the most efficient replicators. Clay-based mineral replicators would have been the first life on earth.

But then why are our feet, not quite literally, made of clay? Cairns-Smith proposed that the charged surfaces of the crystal replicators would have tended to bind organic material such as amino acids or nucleotides. The acquisition of organics may initially have aided the crystal replication process, but eventually there was a *genetic take-over* as the clay's genetic information was passed (for safekeeping) to the more stable molecules of DNA or RNA and their clay bodies were discarded.

The most remarkable aspect of this theory is that it has gained quite a lot of support, despite the fact that there is not a scrap of supporting evidence. One obvious criticism is: where is clay-based life now? There is certainly plenty of clay still on the planet – in environments not very different from those found on the early Earth. No one has yet detected anything remotely resembling Cairns-Smith's clay creatures in all that mud.

Yet, like most origin-of-life theories, it probably does have an element of truth. Clay minerals do absorb organic material onto their surfaces and that absorption can act rather like an enzyme to bring reactants together. This surface binding is particularly effective in promoting poly-merization reactions. James Ferris and Leslie E Orgel of the Salk Institute in San Diego have recently managed to obtain peptide polymers up to fifty-five units long by incubating solutions of amino acids or nucleotides with mineral surfaces.

Günter Wächtershäuser of Munich takes another approach. His start-ing point is the formation of iron pyrites (fool's gold) from hydrogen sulphide and iron. Iron pyrites are certainly common minerals and their formation in sulphide-rich waters generates the electrons that, he pro-poses, could have been harnessed to reduce carbon dioxide to organic compounds. The iron pyrite minerals would also have possessed a charged surface that could bind the organic compounds and promote various reactions leading to the formation of amino acids and nucleotides, rep-licators and eventually – life[11]. Wächtershäuser and colleagues from the Regensburg Institute for Microbiology have demonstrated that the formation of iron pyrites can be coupled to the polymerization of amino acids. However, they have yet to demonstrate that iron pyrite formation can fix carbon dioxide to make organic compounds; so their theory still has some way to go.

What both the clay and the iron pyrites theories have in common is the replacement of the primordial soup with a more structured environment. In both cases this is partly achieved by going from three dimensions to two (the surface of clay or pyrites). The advantage is that many chemically improbable steps, such as the polymerization of RNA bases or amino acids, are made more favourable in these constricted environments.

ISN'T LIFE COMPLICATED ENOUGH?

A relative newcomer to the origin-of-life field is complexity theory. Most people are familiar with its alter ego, chaos theory, in the guise of the butterfly effect, in which a butterfly flapping its wings in an Amazonian rain forest could start a chain of air disturbance eventually causing a hurricane on the other side of the world. The basis for this effect is the extreme sensitivity of chaotic systems, like the weather, to starting conditions. The other side of the chaos coin is that very complex systems can spontaneously generate order rather than chaos: *order for free.*

The weather is a complex system generated by the random movements of trillions of molecules of air and water. Even if the positions and velocities of every one of these trillions of molecules were known at one point in time it would be an impossible task to calculate the state of the weather, even a few moments later. The problem is too complex and the solutions too chaotic. However, any meteorologist could tell you that, at any particular time, there is likely to be an anticyclone (an area of high pressure) over the Azores. Although complex systems, like the weather, have a near-infinite number of possible states, they have a tendency to fall into *attractors*, of simple ordered behaviour. The Azores anticyclone is such an attractor. Despite the near-infinite number of ways that air can travel around the Atlantic, the most stable pattern is a weather system centred over the Azores.

The ability of complex systems to generate order is proposed to be involved in phenomena as diverse as chemistry, meteorology and world economics. The spontaneous order of complex systems has some similarity to life. The Azores anticyclone is an ordered structure that *feeds* on the winds, just as living cells are ordered structures that feed on chemicals in their environment. Both the anticyclone and living cells are dynamic systems rather than static structures; both are continually renewed by the material flowing in and out of them.

There has been a glut of popular books on complexity in recent years, so see the Bibliography for various titles recommended to gain a fuller understanding of this fascinating field. Our interest is the claim made by many complexity researchers that complexity's order for free generated the first living cells. Stuart Kauffman of the Santa Fe Institute is one of complexity's leading proponents. His theory of the origin of life starts with primordial soup containing billions of different kinds of molecules. In such a complex system it is quite likely that some molecule, say A, will catalyse the formation of some other molecule, B. It may also happen that B will catalyse the formation of C which will go on to catalyse D and so on in a series: $A \rightarrow B \rightarrow C \rightarrow D \rightarrow E \rightarrow F \rightarrow G$ etc. However, in such a complex system there is also a possibility that one of the components along the series (say F) will happen also to catalyse the formation of A from the primordial soup, giving *catalytic closure* of the cycle, $A \rightarrow B \rightarrow C \rightarrow D \rightarrow E \rightarrow F \rightarrow A \rightarrow B \rightarrow C \dots$ and so on. The resulting *autocatalytic set* could continually perpetuate itself by feeding on the primordial soup to form a kind of anticyclone of interlocking chemical reactions. The sets could even replicate whenever a few drops of the soup containing one autocatalytic set splashes into another pool to start a new cycle. New chemicals invading a cycle would initiate mutations leading to new and more complex sets. Eventually a genetic take-over could have coupled one of these catalytic sets to RNA or DNA. The autocatalytic set would have become enclosed within membranes and the first living cell was born.

The ability of complex systems to spontaneously generate order is impressive. I remember being mesmerized by one example when I tried it out in my laboratory: the Belosov-Zhabotinski chemical reaction. The reaction is very simple; you mix a few chemicals to make a purple solution in a shallow dish and wait a few minutes. First you see tiny blue dots that grow into a series of circles and waves that soon fill the entire plate.

It seems almost magical that a featureless dish of inky water spontaneously generates these ordered patterns and waves of oscillating colour.

Many aspects of the natural world almost certainly depend on this self-organization. It is probably involved in many aspects of biology, particularly ecology and embryology; but is it capable of generating life? Kauffman and other complexity theorists bolster their ideas by performing computer simulations in which they show that autocatalytic cycles do spontaneously emerge from their digital primordial soups. However, the problem with much of complexity theory is that it is too rooted in this kind of digital simulation and takes little regard of *wet life*. Computer demonstrations of self-organization can be found on hundreds of web-sites but no one has yet managed to find a complex chemical system which spontaneously generates an autocatalytic set. Yet it ought to be easy. Complex chemical systems are generated every time you bake a cake or boil soup. The gunk that forms the predominant product of most primordial soup experiments is a highly complex chemical system. Yet, no autocatalytic sets have (as far as we know) emerged from any of these complex chemical systems. My guess is that the spontaneous emergence of autocatalytic sets is feasible only in computers, where each set can be isolated from the jumble of reactions around them. In real chemical soups, each component gets caught up in a thousand side reactions that inevitably dilute and dissipate any emerging autocatalytic sets.

A second and more important objection to complexity theory, as a theory to explain the phenomenon of life, is that it is not relevant to the generation of ordered structures inside living cells. As Chapter Six explores, the self-organization of either the Belosov-Zhabotinski reaction or the Azores anticyclone is generated by the random interaction of billions of molecules. They are phenomena of huge numbers of particles and have a structure only on a macroscopic scale; at a molecular level there is only chaos and random motion. Yet, as the next chapter will explore further, cells have ordered structures all the way down to funda-mental particles. The macroscopic structures of living cells are not gener-ated by random incoherent motion but by the directed motion of individual particles. Life is a phenomenon of small numbers and must be described by a different set of rules from complexity theory. As Erwin Schrödinger noted in 1944, 'life represents order based on order'. It seems entirely unlikely that a system that generates order through random incoherent motion could have spontaneously given rise to a system – life – that generates order by an entirely different process.

JUST THE CHICKEN

The problem with generating the kind of self-replication that goes on inside living cells is that, at a molecular level, it is not self-replication. DNA does not self-replicate, neither do proteins nor RNA. Living cells have a division of function: enzymes (made of protein) do all the work in the cell including making RNA and DNA. The information for making proteins is stored in DNA; and RNA acts as the mobile messenger of that information. All three components are required for a functional living cell. None can function on their own. But which came first?

This is a classic chicken-and-egg scenario (except it has one chicken and two eggs). If DNA came first, then what made DNA? If RNA came first, then what made RNA? If protein came first, then how was it encoded? A possible solution became apparent when in 1982 Thomas Cech discovered certain RNA molecules could act as enzymes. The first examples were self-splicing introns from a protozoan (single celled animal) called *Tetrahymena*. Introns are sequences that interrupt most genes in eukaryotes. They need to be removed from messenger RNA in a reaction known as RNA splicing. The *Tetrahymena* intron was found to be self-splicing; it could enzymatically splice itself out of messenger RNA. Cech and others went on to demonstrate that short segments of RNA, *ribozymes*, can act as enzymes to catalyse many biochemical reactions.

The discovery that RNA could act as an enzyme was quickly seized upon as a way out of this conundrum. It was already known that many viruses, like the flu virus, have RNA rather than a DNA genome. The discovery of catalytic activity suggested RNA could be both genome and enzyme. Proteins were no longer needed. The first life might have been an RNA molecule, which could act as an enzyme to make itself. An RNA world was envisaged in which different RNA molecules replicated, mutated, competed and were selected by Darwinian evolution. Over time the RNA replicators would have recruited proteins to improve their replication efficiency, along with membranes and eventually DNA (which would have been recruited to act as a more stable repository of genetic information) to generate the first living cell.

The idea of an RNA world of self-replicating RNA molecules that preceded the emergence of DNA and cells is now accepted as almost a dogma in the origin-of-life field. RNA has been shown able to perform the key reactions expected of a self-replicating molecule. Ribozymes can ligate (join up) two RNA molecules. They can also polymerize up to about six activated RNA bases on an RNA template. But nobody has yet designed or discovered a self-replicating RNA molecule.

There are, however, a number of problems with the RNA world. RNA polymerization excludes water and is subject to the same inhibition in water as occurs with proteins. Providing RNA nucleotides that are modified, to make them more reactive, can relieve this, but the chemical step activation is extremely implausible in the primordial setting. Another problem is the chirality of RNA nucleotides. Like amino acid polymerization, ribozyme-catalysed RNA polymerization does not work with a mixture of D and L RNA nucleotide bases. Nobody has come up with a prebiotic mechanism that would enrich one chiral form of RNA bases over another.

An even more fundamental problem is that RNA is completely implausible as a prebiotic chemical. RNA bases are composed of three pieces: the base (the chemical that does the base pairing), a ribose sugar and a phosphate group. All of these together add up to about fifty atoms that have to be put together in a highly specific manner. Although some success has been achieved in synthesizing the bases and phosphate groups in primordial soups, it is far more difficult to generate the ribose sugar. The most credible reaction that does yield some ribose sugars also yields a staggering array of other sugars including ribulose, xylose, arabinose, xylulose, lyxose, allose, altrose, glucose, mannose, gulose, idose, galactose, talose, psicose, fructose, sorbose and tagatose. There is no known prebiotic mechanism that enriches for the ribose sugar. Even if the ribose sugar were made, putting all three components together is an even more formidable task. When plausible prebiotic forms of the three components are reacted together, they combine in all sorts of ways to generate the inevitable gunk that bedevils all primordial soup experiments. Prebiotic chemists often avoid this gunk by using *activated* forms of bases which have various chemical groups added or modified to avoid those unwanted side reactions that lead to gunk. But this is a bit like cheating. The activated bases are, inevitably, vastly more unlikely as prebiotic compounds than even the original RNA bases.

Chemists can and do manage to synthesize the RNA bases from simple chemicals, but do so by going through a very complex series of carefully controlled reactions in which each desired product from one reaction is isolated and purified before taking it on to the next reaction. Cairns-Smith has estimated that there are about one hundred and forty steps that go into the synthesis of an RNA base from simple prebiotic compounds. For each step there are a minimum of six alternative reactions that could each occur rather than the desired reaction. Chemists favour the desired reaction by carefully controlling each step but the prebiotic world would have had to rely on chance. Perhaps the sun came out at just the right time to evaporate a little pool of chemicals; and then it rained and a splash from another pool was mixed in; the sun came out again; a volcano erupted to add a little sulfur, and so on. Cairns-Smith estimated the probability that each of the one hundred and forty necessary reactions would have by chance yielded the right one of six possible products as one chance in 6^{140} or 10^{109}. This is back to the realm of monkeys with typewriters; the Earth simply has not had enough time.

There are many variations of the RNA world that try to circumvent these problems by proposing a pre-RNA world in which some simpler, easier to make, *RNA-like* molecule acted as the first self-replicator. A period of Darwinian evolution of these first self-replicators is proposed to have been followed by a genetic take-over by RNA, resulting in the RNA world, followed by a later DNA take-over, which later gave way to cellular life. We seem to be relying on too many take-overs here.

Another *just the chicken* hypothesis was proposed in 1996, with the discovery that a small peptide could also replicate itself. David Lee and colleagues of California's Scripps Research Institute designed a short peptide, thirty-two amino acids long that could act as an enzyme to stitch two bits of itself together (fifteen and seventeen amino acids long), and replicate[12]. The reaction was however helped along by using *activated* peptide fragments to minimize the gunk-yielding side-reactions. No prebiotic route to these activated amino acids was suggested.

Nevertheless, even if we assume that such a replicator was the ancestor of the proto-cell, it is still hard to imagine how it could have arisen by purely random processes on the early Earth. To illustrate the problem, let us imagine further that there was a protein, perhaps the size of David Lee's thirty-two amino acids peptide, that could replicate itself in the primordial soup. What is the probability that this peptide was generated by a chance combination of amino acids? The probability is easy to calculate, it is 20^{32} (twenty possible amino acids in each of thirty-two positions) or 10^{41}. To put this figure into perspective, the entire mass of organic carbon present in all of today's rain forests is about 10^{15} kilograms. If a similar mass of random thirty-two amino acid peptides were dissolved in the primordial ocean, how many of Lee's peptide molecules would have been present? The answer is zero. In fact it would take five thousand tropical rainforest-sized masses of peptide to be dissolved in the primordial ocean, to have a reasonable chance of finding just one molecule of Lee's peptide in the mixture.

What if the first replicator was an RNA molecule? How easily could this have emerged from the primordial soup? This again is easy to calculate. A typical ribozyme, able to catalyse a simple polymerization step, is about 100 bases long. The probability of its being generated by random assembly of bases is 4^{100} or 10^{60}. Once again, it would take many thousands of rain-forest sized masses of RNA molecules to have a significant chance of making a molecule of the ribozyme molecule. There are also those 140 independent reactions needed to make each RNA base – is it feasible that even one rain-forest sized mass of RNA molecules could have been generated by random inorganic process on the early Earth?

RNA world (or peptide world) researchers justify these unlikely odds by claiming that there are likely to be very many RNA molecules or peptides able to self-replicate. If there are thousands or millions of one-hundred-base RNA replicators then we need much less mass to find one. However, since nobody has yet found a single RNA or peptide molecule that can self-replicate from plausible prebiotic precursors, this claim is somewhat extravagant.

THE ANTHROPIC UNIVERSE

One way out is to invoke what has come to be known as the anthropic principle. This principle was formulated to account for the universe's many peculiarities that seem uniquely suited to the development of life. One of my favourites concerns the peculiar properties of water. You may recall that water molecules tend to form aggregates joined by weak hydrogen bonds. The exceptional ability of water to form these bonds is critical to biochemistry. However, it depends on *coincidence* – that the bond angles between the oxygen and hydrogen atoms allow water molecules to form near perfect tetrahedrons. This coincidence is why ice floats on water. If ice were like any other solid it would be heavier than water and fall to the ocean bottom. In a *heavy ice world*, whenever the atmosphere dropped below freezing, the warmer surface water would freeze and sink. Convection currents would rapidly transport warmer waters to the surface and dump the colder waters, as ice, into the depths. Soon the entire ocean would freeze, from the bottom up. Aquatic creatures would become entombed in ice.

Real water is heavier than ice so none of this happens. Instead, when water freezes, it floats to the ocean surface, forming a thick insulating layer that protects the warmer water below. Life continues to thrive below the ice. If it weren't for ice's insulating layer, Earth's oceans would be far more inhospitable places, particularly during its many extended Ice Ages. Although microbes could probably have survived prolonged global ocean freezing, higher forms of animals would have been subject to frequent catastrophic extinctions that would have severely slowed the pace of evolution.

We therefore owe our place on this planet to a peculiarity of the structure of the water molecule, dependent upon the electronic structures of hydrogen and oxygen atoms. These atomic structures depend in their turn upon the precise values of two fundamental constants[13]. Change these constants only very slightly and ice would sink, hydrogen bonds wouldn't form and complex life would never have emerged on Earth (or probably anywhere else). Why the fundamental constants have the particular values we observe is completely unknown.

In 1973, Brandon Carter suggested that these and other coincidences are accounted for by our own existence. If the values of the fundamental constants were just different enough to rule out the possibility of carbon-based life then we would not be here to muse on the unfortunate near miss. The universe is anthropic in the sense that its properties must be compatible with the emergence of life and indeed intelligence. There may be (or have been) billions of alternative universes, with different values for their fundamental constants, but they are probably all sterile. The one fact that we know for certain about our universe is that its properties are compatible with our own existence.[14]

One coincidence necessary for the emergence of intelligent life was the sequence of prebiotic chemical reactions that gave rise to the first living cell. In this sense, we do not need to account for the improbability of those events. According to the anthropic principle, we are here so they must have happened. If prebiotic chemistry had not travelled along the route that led to the first self-replicator, then we would not be here to worry about it. We may be the end result of a bizarre chemical accident that generated the first living cell. Since life on Earth is the only life we know of, it doesn't really matter how improbable that was. Its emergence might have been an extraordinarily improbable event – unique in the entire universe or even in a multitude of universes. The anthropic principle tells us, however improbable the event was, it must have happened once and we are its descendants.

You cannot argue with the logic of the anthropic principle but I must admit to profound dissatisfaction with it. That life is so fantastically improbable that we are likely to represent its only example in the entire universe, is deeply depressing. What was the point of watching all those *Star Trek* episodes or *Alien* films if there is nothing out there! Admittedly, there is currently no evidence for extraterrestrial life, apart from some peculiar shapes in Martian meteorites. But I would like to think that when probes are sent beneath the ice on Europa, they will find life. If extraterrestrial life is ever discovered then (presuming it originated independently), the anthropic principle, as an explanation of life on Earth, will be dead.

Unfortunately, hopes are far from scientific evidence. In the absence of evidence of extraterrestrial life, it is hard to argue against the anthropic principle. I do however have a couple of counter-arguments. The first is that life's emergence was quick. As far as we can tell, life emerged on Earth as soon as it was physically possible (when liquid water was stable on the surface). This suggests that, when conditions were right, life was not only possible, but probable. The second point is that I do not believe that we need the anthropic principle to account for the emergence of life. As the next chapter will show, life is governed by different kinds of rules from those than the randomizing forces that drive inanimate chemistry. I hope to persuade you that these rules – quantum mechanical rules – give us another way to overcome the huge improbability of the first self-replicator.

5

Life's Actions

After exploring the limits of life, our alien spacecraft might next want to find out how life works on Earth. Its next dispatch might run something along these lines: *twenty-two cylindrical carbon/water-based bodies of mass eighty to one hundred kilograms performed a series of actions, accelerating from rest to velocities of one to five metres per second, changing direction and speed and decelerating at irregular intervals. Their actions appeared to be independent as no external forces were detected. Many of their actions impinged upon a smaller spherical object made of the same materials as the larger bodies but with a mass of four hundred and forty grams. In contrast to the movement of the larger objects, this body was only observed to move when under the influence of applied external forces.*

Our inquisitive alien spaceship might have detected a human action, a game of football, and spotted the essential difference between the players and ball. The ball moves only when forced; whereas the player's movements are independent of external forces – they are *independent actions*. The makers of the spacecraft would, like us, be curious to discover the source of these actions. This chapter will attempt to peel back the layers of mechanisms behind the independent actions performed by living creatures. We will witness the success of reductionism in providing an explanation of life's actions at one level; but also discover that when trying to dig deeper, the reductionist approach paradoxically points us in a surprising new direction.

MUSCLE POWER AND THE MYOSIN RATCHET

So what happens when you kick a ball? First, *you* must decide that kicking the ball is a good thing to do. Kicking is nearly always a voluntary action and, as such, is initiated by an electrical impulse in your brain. Later, we will explore the source of that electrical impulse, now, let us concern ourselves with what appears to be a much simpler question: the mechanics of the interaction between the ball and your body. What makes the ball move? The simple answer is its impact with your boot. And what makes your boot move? Your leg, of course. And what makes your leg move? The contraction of muscles inside your leg. But what makes muscles contract?

Muscles are engines that convert chemical energy into mechanical work. We are all reasonably familiar with car engines where chemical energy is also converted to mechanical work. In a typical engine, the (petroleum) fuel is first vaporized, then injected or blown into the engine cylinder. A spark ignites the gas mixture and generates a great deal of heat, expanding the cylinder gases to push the pistons that turn the engine. The source of a car's power is the chemical energy stored in petrol. But what about the force that drives the foot that pushes the accelerator? What is the source of that power?

A protein called *myosin* moves muscles. Myosin is a very big protein made up of several thousand amino acid beads. To understand how it works, we need to understand the shape of proteins in living cells. Although proteins are described as strings of amino acid, they are not like floppy strings of beads but more rigid folded structures (think of beads on wire not string). Each protein wire has many loops and turns that are dictated by the interactions of amino acids, both with themselves and the surrounding water molecules. These interactions force each protein to adopt a characteristic three-dimensional shape that is key to its function. The shape of the myosin molecule is a globular head connected to a long tail by a flexible hinge region. It is the flexing of this hinge that provides the energy for muscle contraction.

Muscle fibres are packed with thick filaments made of myosin and parallel thin filaments mainly made of another protein called actin. Each myosin tail is anchored within the thick filament. The globular heads are reversibly attached to the actin filament to form a dense series of cross-bridges between the thin and thick filaments. Actin filaments are relatively rigid structures in animal cells that act like tramlines, along which organelles, chromosomes and myosin heads are transported. During muscle contraction, the myosin hinge region goes through cycles of flexing and relaxation that transposes the head protein, in ten-nanometre steps (a nanometre is one billionth of a metre), along the actin tramline. Thousands of myosin molecules act like an army of motorized ratchets to pull the myosin fibres along the actin tramlines and thereby contract the muscle fibre.

But what drives the cycle of flexing and relaxation within the myosin molecule? The key to the protein's dynamics is its ability to act as an enzyme, to promote a reaction between an ATP (adenosine triphosphate) molecule and water. The reaction, called hydrolysis, breaks the ATP molecule into two smaller pieces and the myosin enzyme captures some of its chemical energy to drive its cycle of flexing and relaxation. To see how the enzyme manages this, we need to know a little about chemical reactions in general and how enzymes promote certain reactions. If already familiar with chemistry and biochemistry, you might want to skip the next few sections; but for those who aren't, I will give you a quick – and hopefully painless – tour of these sciences.

A CHEMICAL INTERLUDE

Chemical reactions always involve changes in molecular structures: to understand them, it helps to know how matter is structured. One of the most straightforward materials is common table salt, which is made up of only two elements: sodium and chlorine. Although sodium is solid at room temperature, chlorine is normally a gas. The reason that the chlorine atoms do not simply float out of salt is that they are tethered within salt crystals by the energy of chemical bonds.

The simplistic view of atoms is of negatively charged electrons orbiting the nucleus with its positively charged protons and neutral neutrons. The constituents of the nucleus in chemistry are not involved in chemical change (they are changed in nuclear reactions such as the fission reactions that power atomic bombs). But all chemistry involves electrons moving in one way or another; so to understand the chemistry of life we need to know a little about electrons. Fortunately, electrons are completely structureless fundamental particles. Know one electron, know them all! Each and every electron has precisely the same mass and carries precisely the same electric charge as every other electron. Electrons differ in only two properties: their *spin* (which can be either clockwise or anticlockwise) and their energy. Electron spin need not concern us here; for now, we need only consider electron energy.

Energy comes in two forms – *kinetic* and *potential*. Kinetic is the energy inherent in moving bodies. Think of a flowing river. The water in a river possesses the kinetic energy that moves rocks, stones and sediment along its path. Build a water mill in its course and this energy can be captured and used to grind grain. Potential energy is subtler. It is the energy inherent in bodies due to their position in an energy field. Water in a mountain lake possesses potential energy due to its position within the gravitational field that surrounds our planet. The energy is potential, rather than immediately manifest, but it can be harvested, for instance, by channelling the water downhill. As the water flows, its potential energy is converted to the kinetic energy of a river or waterfall. That kinetic energy will eventually be dissipated both as heat and through the mechanical forces that roll stones or carve a pool out of the rock below a waterfall. The key point to bear in mind is that energy fields (such as Earth's gravity) tend to move bodies from places of high potential energy to places of lower and in the process the difference in energy may be released or captured. In physics this potential to make a body move is termed a force. Earth's gravity field exerts a force on water molecules tumbling down a hillside. Contracting muscles exert a force on our limbs. The atoms of chemical compounds, like salt, are held together by force. The myosin molecule exerts a force on the actin filament, to make it move.

The next chapter will look more closely at what forces are. However, whenever a force makes a body move (or more accurately, causes a change in momentum), then in physics the event is described as an action. No action is performed when a body continues to move at the same speed and direction as it is already moving, since (in the absence of frictional forces), no force is required for that movement. An action only occurs when a force is involved.

The force that moves myosin, muscle, people and cars and holds the atoms together in salt, is the electromagnetic force, or simply em (pronounced *eeee-em*) force. Charged particles, such as electrons or protons, generate the em force. The movement of electrons in a metal wire generates an electric force that can drive a motor. The rotation of electrons around atoms within a bar magnet generates the magnetic force. Magnetic and electric fields are two manifestations of the same phenomenon, electromagnetism. Chemical energy is also a manifestation of em force and it is generated by the interactions between charged protons and electrons within matter.

If a negatively charged electron is brought close to another negatively charged particle (such as another electron), the em force will repel the pair (like charges repel) like a compressed spring. If the first electron is released, its potential energy (due to its position within an electromagnetic field) will be instantly converted to kinetic energy and, just like a compressed spring, it will speed off into the distance. Conversely, if an electron is brought close to a positively charged particle such as a proton, then the em force will pull the electron and proton together (like a stretched spring). The electron, being the lighter of the two particles, will tend to move towards the proton. Why electrons in atoms do not, in fact, spin down into the protons in the atomic nucleus is a profound question that requires quantum mechanics to answer it. For now we will have to content ourselves with the assurance that if an electron gets too close to a proton, then their combined potential energy increases. So a free electron will tend to fall towards a proton until it is trapped within a well of minimum potential energy. Pull the electron and nucleus further apart and attractive em forces will pull them back together. Push them together and they will resist by pushing apart.

The em force is very strong. Bulky objects like ourselves do not normally feel its manifestations, because on this scale all the positive and negative charges within atoms and molecules of matter tend to cancel each other out, leaving zero net charge and zero net force. Only when we separate the tiniest fraction of these charged particles in, for instance, a battery, do we normally notice the effects of the em force. But if all the positives and negatives in matter did not cancel each other out, then the full strength of this force would be enormous. Imagine if it were possible to separate all the positive and negative charges in two grains of salt, putting all the positive charges in one and all the negative charges in the other; if the, by now extremely positive, grain of salt, were placed thirty metres away from the extremely negative grain of salt, then the total em force between the grains would be strong enough to haul a load weighing three million tonnes.

In bulk matter, nearly all the attractive and repulsive forces cancel out to leave net em force of zero between any two (uncharged) objects. But on the scale of atoms and molecules, the em forces do not cancel. Electrons feel the pull of protons. Protons feel the pull of electrons. All chemical reactions involve a shuffling, pulling and pushing of electrons and protons within matter to find the arrangement that has the lowest potential energy. Even within atoms, electrons shuffle around to find their lowest energy. Electrons are said to 'orbit' the nucleus in a series of electron clouds or shells (although they don't really orbit since they do not really have any kind of defined spatial position that would allow such orbiting – see later chapters on quantum mechanics). The simplest atom, hydrogen, has only a single positively charged proton in its nucleus and a single negatively charged electron in the first electron shell. Helium has two protons and two electrons. Lithium has three protons and three electrons (all atoms have the same number of electrons as protons, to maintain electrical neutrality) and so on. Each electron shell is like a room that can only house a certain number of electrons. The lowest shell has room for just two electrons whereas most higher shells can accommodate up to eight. Electrons in higher shells/rooms experience an em force (from the field of the protons) that pulls them down into the lowest unoccupied space in a room where their potential energy is at a minimum. Chemical reactions involve the movement of electrons from high electron rooms in one atom to low electron rooms in another.

The hydrogen atom with one electron room occupied by just one electron can accommodate one guest electron in its single room. This makes it quite reactive. Helium already has two so it has a full house. The contented helium electrons are rather unsociable – they do not welcome guest electrons and they rarely leave their home – so the element is very unreactive. Lithium has three electrons, two in its lowest room and one electron in its outer electron room. The lonely outermost electron is keen to find company, so is easily persuaded to leave home, making lithium very reactive. The next element is beryllium with four electrons, which tends to lose both its top electrons. Boron has five electrons, carbon six, nitrogen seven, oxygen eight, fluorine nine and neon ten. Neon has eight electrons in its outer shell and so a full house and, like helium, is unreactive. The element below neon is fluorine, which has just seven outer electrons in its outer shell. It happily accepts a guest electron to fill its outer shell and achieve a full house. Similarly oxygen, with six outer electrons, is happy to accept a pair of guest electrons.

Chemical reactions involve the movement of electrons from one atom to another, down a potential energy gradient. The formation of common salt (sodium chloride) illustrates this process. Sodium is a soft extremely reactive metal which ignites spontaneously on contact with water. Chlorine is the noxious yellow gas released from household bleach. Because it is so reactive, it is highly toxic and can cause severe lung damage if inhaled in sufficient quantity. Yet the union of these two unpleasant chemicals is the entirely innocuous, sodium chloride. Each atom of sodium has a total of eleven electrons. Two electrons are in the inner electron room, eight in the next and only one in its top electron room. This outer electron has a relatively high potential energy. The chloride atom has seven electrons in its outermost room and therefore has a vacancy for a single electron. If a sodium atom is brought into contact with a chlorine atom, there will be a gradient of electromagnetic energy between the two atoms – high on the sodium side, low on the chloride side. The outermost electron in the sodium atom will experience an em force that will cause it to move electromagnetically downhill. It will travel from the sodium electron room to the vacant space in the chlorine atom. In doing so, it will release the difference in its potential energy as radiation (light) or heat. The chlorine atom (symbol Cl) will acquire an extra electron to become a negatively charged chloride *ion* (symbol Cl^-; ions are atoms that have lost or gained negatively charged electrons and so have a net charge). It now has the optimal eight electrons in its outer shell. The sodium atom (symbol Na) will have lost an electron. What is now its top electron shell already possessed the optimal eight electrons; so it is also content. The sodium atom has lost an electron so it has a surplus of a single positive charge and thereby becomes a positively charged sodium ion (Na^+).

An action has taken place during this chemical change. The action involved the movement of an electron from the sodium atom to the chlorine atom. The force involved was the em force between the two atoms. All chemical reactions involve the movement of electrons in response to the em force and are therefore all *actions*.

Now that the atoms of sodium and chlorine have become oppositely charged ions, the em force between them is attractive and forms a chemical bond. Pairs of oppositely charged sodium and chloride ions are pulled together to form the solid crystalline lattice of sodium chloride (NaCl) or common salt. The bond between the sodium and chloride ions is known as an ionic bond. It is these ionic bonds which prevent chlorine atoms from floating out of salt crystals. You break thousands of these bonds whenever you grind salt in a salt mill. The force needed to do so may not seem particularly impressive but that is because you are breaking only a tiny proportion of the billions of ionic bonds in a salt crystal.

Another very important chemical bond is the type found in water. Atomic oxygen has six electrons in its outer electron room – two short of the optimal eight – so is able to accommodate two more electrons. Hydrogen atoms have only a single electron. Oxygen atoms can acquire two electrons, one from each of two hydrogen atoms. However, instead of each hydrogen atom simply donating the two electrons, the atoms share a single electron room that houses all eight. When the hydrogen and oxygen atoms are brought into close enough proximity, the electrons from each hydrogen atom migrate downhill in their potential energy to occupy the new electron shell orbiting both the hydrogen and oxygen atoms, forming the water molecule, H_2O. The outer electron shell of water orbits both the oxygen atom and the hydrogen nuclei to form what is know as a covalent bond. Covalent bonds are very important in biology because most biochemicals like DNA, proteins and fats consist mostly of molecules with atoms held together by covalent bonds. Note that the formation of a covalent bond still involves an action. The electrons that used to orbit the single proton in the hydrogen atom now have a quite a different orbit – they have moved.

WATER

The last chapter showed just how important water was for life. Why? As described above, water is a covalent molecule but also possesses a vestige of ionic character. A water molecule is asymmetric; the oxygen atom at one end, the two hydrogen atoms at the other. Its oxygen atom tends to pull the electronegative electrons to its end to generate a slight negative charge; leaving the hydrogen end with a slight positive charge. This electrical dipole is responsible for many of the unusual properties of water. Each (slightly) positively charged hydrogen atom can behave as an ion and thereby form a (relatively weak) ionic bond with the (slightly) negatively charged oxygen atom of another water molecule: a hydrogen bond. In ice, the water molecules are held together in a regular three-dimensional crystalline lattice. In the liquid state, water molecules are also held together by hydrogen bonds but the bonds are more fluid and dynamic and pack tighter together. This is why solid ice floats on liquid water. Hydrogen bonds also form between water and the chemicals dissolved in it. The solubility of proteins depends on an extensive network of hydrogen bonding between water and charged groups (ions) on the surface of proteins.

Water performs another ionic trick by dissociating into a positively charged H^+ hydrogen ion and a negatively charged hydroxide, OH^- ion. Pure water does not dissociate very much; in fact only one in about ten million[1] water molecules is ionized. Because each water molecule dissociates into a single hydrogen ion and a single hydroxide ion, pure water contains an equal number of both of these ions and has a neutral pH. Acids are chemicals that generate an excess of hydrogen ions in solution and alkalis or bases generate an excess of hydroxide ions in water. pH is a measure of the concentration of hydrogen ions in solutions.

One final property of water molecule, involved in many biochemical reactions, is its ability to either donate or accept a pair of electrons. Water molecules thereby serve as highly mobile electron carriers inside living cells, accepting or donating as required.

It should be clear now why water is so important for life and why most biologists are sceptical of the notion of life based on any other medium. No other chemical has anything like water's extraordinary range of activities.

Now armed with our chemical crash course, let us return to the question of just how muscle myosin works. We left off observing how myosin acts as an enzyme to promote ATP's hydrolysis, capturing some of its chemical energy to drive muscular contraction. The next step in unravelling the process must be to discover how enzymes work.

ENZYMES AS MOUSETRAPS

Enzymes are key ingredients of life, ubiquitous in living tissue. Moisten some fresh bread with warm water and then leave it for about ten minutes. Pop the same amount in your mouth, chewing it for a minute before taking it out, leaving it alongside the other unchewed bread. After ten minutes, taste the unadulterated bread and the bread already chewed (not so disgusting – it was in your mouth just before). The first (just wetted with water) will taste like soggy bread. However, the bread that was in your mouth will taste sweet. The sweetness is due to the presence of sugars released from starch in the bread by the action of the enzyme *amylase* in your saliva.

Starch is a polymer of glucose manufactured by plant cells as an energy store for their seeds. It will hydrolyse (react with water) spontaneously to release glucose, but the reaction is normally very slow. Enzymes are catalysts that speed the rate of chemical reactions by directing the motion of electrons and protons along specific paths. The enzyme in our saliva that breaks down starch is called amylase. Chewing bread impregnates it with amylase and so stimulates the breakdown of starch more than a millionfold.

Your body has the capacity to make thousands of distinct enzymes, each with a specific role to accelerate at least one of the millions of everyday chemical reactions that take place in your body. To discover how enzymes work, we will examine one in more detail. If, like me, you are relatively unfit, you will experience breathlessness, tiredness and possibly even cramp as you run the length of a football pitch. This is due to oxygen starvation in your muscles. When muscle cells have plenty of oxygen they burn glucose completely to carbon dioxide and water (this metabolic burning – respiration – will be discussed more fully below) and the energy released is used to drive muscle contraction. However a plentiful supply of oxygen is required for respiration. During vigorous exercise, the blood supply cannot keep up with the oxygen demand, so the glucose is only incompletely broken down into a chemical called lactic acid. The muscle weakness and cramp you might suffer during vigorous exercise are due to the build-up of lactic acid in your muscles. The breakdown of glucose to lactic acid involves a chain of chemical reactions; each accelerated by its own enzyme. The last enzyme in the chain, lactate dehydrogenase (LDH) converts pyruvate to lactic acid. Like most biochemical reactions, it proceeds extremely slowly without a catalyst. LDH accelerates the rate of the reaction more than a billionfold.

As with all chemical reactions, the reaction catalysed by LDH involves the migration of electrons; in this case, the transfer of a pair of electrons from a molecule (NADH) to pyruvate to yield lactic acid. Enzymatic reactions that involve the donation or removal of a pair of electrons usually involve NADH, which acts as an electron carrier in the cell. After donating its pair of negatively charged electrons, NADH becomes the positively charged ion, NAD^+.

A great deal is known about the reaction catalysed by LDH because the enzyme has been crystallized. The crystallization process aligns all the protein molecules in the same direction within the crystal. In X-ray crystallography, X-rays can then be shone through the crystal to obtain an X-ray picture of the molecule. Scientists studying this X-ray pattern can then map the positions of each amino acid and charged groups within the protein to discover how the enzyme works.

The enzyme can be very schematically represented as a mousetrap. The substrates (NADH and pyruvate) float into a cavity inside the enzyme – the active site. A positively charged amino acid (arginine), at position 171 in the protein (171st amino acid from the left end of protein) serves to anchor the lactic acid; while another arginine at position 101 serves to anchor the NADH. These anchoring groups act as the trap's trigger, releasing a sequence of shape changes within LDH that folds the protein over the substrates and swings an armoury of charged amino acids to attack their chemical bonds. The carbonyl bond (C=O) of pyruvate is very electronegative and thus vulnerable to attack. A positively charged histidine amino acid at position 195 of the enzyme delivers the *coup de grâce*, spitting a proton into this carbonyl group to form a new covalent bond, the hydroxyl (O-H) group of lactic acid. The extra positive charge (from the proton) acquired by the substrate sets up a wave of electron migration throughout the molecule that draws a pair of electrons plus a proton from the NADH cofactor, to yield the products of NAD$^+$ and lactic acid.

After the LDH enzyme has performed its job on pyruvate, it plucks a proton from a water molecule to replace the one lost. This returns the enzyme to its starting state and readies it to attack another pyruvate molecule. A vital principle of all enzymology is that the enzyme (or indeed any catalyst) must emerge unscathed from its chemical gymnastics. Only in this way could a single molecule of enzyme convert millions of molecules of substrate into product. It is this facility to be reused that makes enzymes so effective.

Thus, the key to the action of LDH, and indeed all enzymes, is the directed movement of protons and electrons within the active site of the enzyme. The dynamic interaction between the substrates and enzyme directs electrons and protons along the paths that lead to the products. Enzymes perform directed actions.

BACK TO MUSCLES

So that is how a well-characterized enzyme works; now we can return to examine how the myosin enzyme drives muscle contraction. One player – and a vital one for all cellular life – is ATP which acts as the cell's mobile energy store. The molecule acts as a tiny battery to store chemical energy in the cell. Whenever ATP is hydrolysed (reacts with water) this energy is released and can be harnessed to drive chemical reactions or perform physical work. In muscles, its chemical energy is transformed into mechanical work but it is also used for most of the energy-requiring reactions of living cells.

The key to unravelling the mechanism of myosin contraction was the crystallization of fragments of the myosin head protein by a University of Wisconsin group led by Ivan Rayment. Myosin is a much bigger

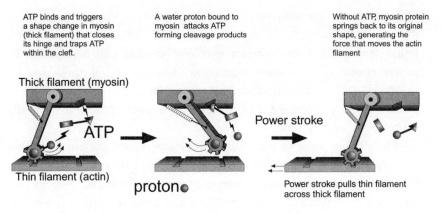

ATP binds and triggers a shape change in myosin (thick filament) that closes its hinge and traps ATP within the cleft.

A water proton bound to myosin attacks ATP forming cleavage products

Without ATP, myosin protein springs back to its original shape, generating the force that moves the actin filament

Thick filament (myosin)

ATP

Power stroke

Thin filament (actin)

proton●

Power stroke pulls thin filament across thick filament

FIG 5.1   The enzymatic action of myosin that is used to power muscles.

molecule than LDH so its crystallization proved extremely difficult. However, in 1993, Rayment and his colleagues finally succeeded and were able to determine the structure of the myosin protein by X-ray diffraction studies. Many aspects of the action of myosin remain unclear but they have proposed a model[2] that fits all the known facts.

Their model begins with each myosin head protein attached to actin. An ATP molecule drifts into the hinge-like cleft between the head and tail of the protein, where it becomes bound (FIGURE 5.1). The electrons and protons in the ATP molecule attract charged groups in myosin, closing the hinge and trapping ATP in the cleft. The protein backbone is put under mechanical stress by this action, akin to the stretching of a coiled spring. This shape-change breaks the interaction between myosin and actin. It also swings a water molecule – bonded to a charged amino acid on myosin – to attack ATP's final phosphate bond. The bound water molecule acts as a proton popgun, firing a single proton into ATP, breaking it into two halves: ADP and phosphate. This hydrolysis reaction induces a second shape-change inside myosin, allowing it to bind to actin again and forcing the ejection of the cleaved phosphate group. In the wake of the phosphate group, the strong electromagnetic interactions holding the hinge closed are lost. The hinge springs open, swinging the myosin head back into its original configuration. However, while the hinge contraction took place, with the myosin head hanging loose, the spring-back occurs with the myosin head protein firmly anchored to the thin actin filament. This myosin head molecule thereby acts as a lever, pulling thick and thin filaments over one another and generating the power stroke of muscular contraction. A sequence of these (ATP-mediated) cycles of contraction and relaxation causes the myosin head protein to creep along the actin filament in a series of steps that rapidly contract the muscle fibre.

The driving force of muscular contraction is thus the em force that distorts the myosin protein during its interaction with ATP. This is how the chemical energy of ATP is harvested – its electrons and protons pull and push the myosin molecule to open and close its hinge. The whole process is mediated by the myosin enzyme *directing the movement of a proton from a water molecule to ATP*. This is the directed action that initiates muscle contraction.

WHERE DO WE GET THAT ENERGY FROM?

It is hardly surprising that the motion of protons and electrons drives our limb's motion. Perhaps more surprising is that the directed motion of fundamental particles drives all life's actions. Consider, for instance, the reactions that are used to make ATP, that mobile energy store in the cell. Each of us makes and consumes about roughly two and a third kilos of ATP every day in our bodies. Where does it all come from?

Animals obtain energy from food. Food serves as both as a source of energy and of raw material from which to construct new cells or repair old ones. We are what we eat. Say you are playing football and you are about to sprint into action. Millions of myosin molecules in your calf and thigh muscles are poised to start their cycle of grasping and releasing the actin filaments, but there is a sudden crisis – they are out of ATP! You can't run to the local store to grab a jar of ATP (to provide the instant power of Popeye's can of spinach) because you cannot even run without ATP. Instead, your body must break down glucose (or other sugars) from your food to make ATP. Although glucose holds plenty of chemical energy, it cannot be used directly for muscle contraction. So muscle cells must transfer incoming glucose supplies into the more readily utilizable form of ATP. Glucose contains six covalently bonded carbon atoms tied into a ring structure: with plentiful oxygen, cells are able to oxidize completely, one glucose molecule to yield just carbon dioxide, water and about thirty ATP molecules.

This brings us to another vitally important chemical reaction: oxidation. Oxidation occurs when you burn paper, wood or glucose in air and it involves electron motion. Electrons in the relatively high-potential energy rooms (orbitals) in the atoms of paper, wood or glucose, travel down a potential energy gradient to fill oxygen's empty low-potential energy electron rooms. The difference in potential energy is emitted as the light and heat of a fire. The energy released from burning wood can be harnessed to generate motion: steam engines power the wheels of a train and in the process, chemical potential energy is converted into kinetic energy. Similarly, cells burn metabolic fuels in an oxidation reaction that yields chemical energy in the form of ATP. The reaction is known as respiration and, just like burning, it usually requires oxygen.[3] The efficiency of our cellular engines is pretty good, about thirty-eight per cent of the total energy available from glucose oxidation is captured in ATP. The rest is dissipated mostly as heat – which is why vigorous exercise makes you hot.

Respiration takes place inside cellular organelles (the organs of cells – the structures doing specific tasks) known as *mitochondria*. The structure of mitochondria is the first indication of respiration's remarkable nature. They have many of the features of whole cells, such as internal membranes, ribosomes and their own DNA. Mitochondria are even able to divide independently from the cell. It is now thought that (like plant chloroplasts) mitochondria are direct descendants of symbiotic bacteria.

In respiratory oxidation, the electrons from glucose are first dumped onto the NAD electron carrier molecule mentioned above, to form NADH. The electrons in NADH – still at high-potential energy – are then plugged into mitochondria, where they are passed – like a relay baton – through a series of enzymes carrying them down the potential energy gradient towards oxygen. Mitochondria are bounded by a pair of membranes – one inside the other – with a water-filled space inbetween; and it is within these membranes that all the electron transport action takes place. Respiratory enzymes, inserted into the inner membrane, act as electron-powered proton pumps, capturing protons from the mitochondria's inside and transporting them into the water-filled space between the membranes. One proton-pumping station is an enzyme (cytochrome oxidase), whose position in the electron relay is to receive electrons from another enzyme (cytochrome c), passing them to oxygen. Buried within cytochrome oxidase are three copper atoms which pluck the electrons from an iron atom within the cytochrome c and then push them on to oxygen. Inserted into the inner membrane, one end of the protein faces towards the mitochondrial matrix on the inside of the mitochondrion, and the other end faces the intermembrane space. On receiving electrons, the enzyme undergoes a shape-change, causing a charged amino acid on the matrix side to pluck a proton from a water molecule within the matrix. A second shape-change causes another charged amino acid to spit a proton into the intermembrane space. The result is an electron-flow-driven pumping of positively charged protons from one side of the inner membrane to the other, generating an electrical charge difference about 0.15 volts – in other words, a battery.

The mitochondrial battery then powers ATP synthesis. The enzyme involved is called ATP synthase or ATPase and acts as a molecular motor. The enzyme spans the inner mitochondrial membrane so that it experiences the proton gradient generated by the respiratory enzymes. The enzyme faces more than ten times the number of protons at its head than at its tail. A channel in the protein allows protons to flow through. This proton flow causes the enzyme to rotate – rather like a turbine engine or water wheel. Just as the energy of a turning water wheel is harnessed to grind corn, so the energy of the rotating protein is used to synthesize ATP. The enzyme rotates in steps of 120 degrees and each step is associated with the synthesis of a single molecule of ATP.[4]

The precise mechanism by which ATP synthase is coupled to ATP synthesis still remains unclear. A reasonable conjecture is that the rotation twists the protein structure so that, like a coiled spring, it can be made to do work. Quite likely the shape-changes within the protein bring about charge migrations that make ATP. However, such a remarkable enzyme may have even more surprises up its sleeve . . .

In any event, enough is known about respiratory ATP synthesis to be certain that it is one of the universe's most remarkable phenomena. We wonder at our engineering feats – the jet aircraft, Channel Tunnel and the microchip, but these pale beside the marvels of those tiny electric-powered proton pumps and turbine engines that power every living cell inside our bodies.

The ATP energy driving our muscles is derived from the chemical energy of food molecules, such as glucose. That energy's ultimate source is sunlight harnessed by plants. Plants are notoriously immobile (apart from a few famous exceptions such as the Venus Flytrap). Surely they don't move when they perform photosynthesis? But they do.

Plants and those microbes that perform photosynthesis capture the energy of light photons (particles of light) to make ATP. The reaction takes place within chloroplasts inside plant cells. Chloroplasts are much like mitochondria with internal membranes loaded with chlorophyll where most of the action takes place. Like mitochondria, they use an electron-driven proton pump to generate a proton gradient that drives an ATPase turbine engine. However, whereas mitochondria derive their electrons from NADH, chloroplasts capture electrons directly from water. But water electrons do not possess sufficient energy to drive the proton pump – this is where light comes in.

Plant-pigment molecules capture photons of light, funnelling their energy towards a magnesium atom in the centre of a chlorophyll molecule. The chlorophyll's outermost electron absorbs the photon's energy, energizing it, and causing it to pop out of its atom. This energized electron is then used to drive a membrane proton pump. As with mitochondria, the resulting proton gradient turns an ATPase turbine engine to make ATP. After losing an electron, the electron-depleted magnesium atom is sufficiently *electron-hungry* to pluck an electron from a water molecule, generating a by-product of oxygen – which is why plants excrete oxygen.

The essential reaction in photosynthesis is, once again, the directed motion of a fundamental particle – in this case an electron. The force that moves this electron is captured from the absorption of a light photon. But it is the plant's enzymes which *direct* that motion, thereby harnessing its energy, to make ATP.

So Aristotle was right; all life involves movement. The movement of animals is obvious, but movement is equally important for stationary plants – the only difference is that they perform most of their motion on a microscopic level. At the level of the living cell, all biological activity – growth, photosynthesis, digestion, metabolism and replication – involves the *directed movement of fundamental particles*. This is why life is so different from the chaos-engendered order of an anticyclone or the Belosov-Zhabotinski chemical reaction described in the last chapter. Chaotic motion lies at the heart of those phenomena. But the motion inside living cells is ordered – all the way down to the level of fundamental particles: order from order.

But isn't life now fully explained? The realization that the directed motion of fundamental particles underscores all the actions of living cells seems to leave us with nothing else to account for. What does the rest of this book deal with? Well, in fact, we haven't yet accounted for life at all. We have only recognized that motion lies behind the phenomenon. But what causes that motion? How is it directed? So far all motion has been explained in terms of *electromagnetic forces*. But what is an electromagnetic force? What, indeed, is any force? Or more generally, what makes things move?

6

What Makes Bodies Move?

Living organisms are not, of course, the only moving bodies in the universe. Trains, cars, rivers, planets and entire galaxies all move. What makes the movement inside living cells so special? To answer this, we must first ask why anything moves. When we kick a ball hard enough it soars up into the air. We have just examined the muscular basis for the swinging of our booted foot; but what happens when it impacts with the ball. Why does the ball move? The simple answer is that the boot pushes it. But what do we mean by this? My naïve assumption might be that our swinging boot travelling at a certain velocity contacts the stationary ball, continuing its movement and thereby pushing the ball along until the ball gains enough velocity to travel of its own accord. One problem with this scenario is that our boot and the ball never actually contact each other. As the atoms and molecules of our boot and the ball are brought closer and closer, they do not ever reach a point during the kick when they actually touch. What then is doing the *pushing*?

We need to look more deeply at movement, and for that we need the science of movement: mechanics. Aristotle wrote at great length about movement, but was mostly wrong. He believed, for instance, that heavy objects fall faster than lighter ones. The modern science of mechanics owes its origin to Galileo's experiments to prove Aristotle wrong.

Galileo Galiliei was born in the same year as Shakespeare, 1564, in Pisa, Italy. He studied medicine at Pisa University, then turning to mathematics and moving to a chair at Padua University where he remained until 1610. During these years Galileo constructed many of his most brilliant inventions including a thermoscope (a kind of thermometer), a compass, a telescope and a hydraulic pump. Many of these inventions were measuring devices for experiments in astronomy and dynamics and reflect what was actually Galileo's greatest innovation: his insistence that scientific knowledge must be gained by empirical observation. This went contrary to prevailing opinion, which generally regarded the writings of the ancients as the ultimate authority. According to Francis Bacon, if two scholars disagreed over the number of teeth in a horse's jaw they would consult Aristotle rather than look into the animal's mouth. Galileo was the first modern scientist who denied Aristotle's dogmas, designing experiments to test his ideas. He constructed ingenious measuring instruments, such as a clock that measured time elapsed by the rate water poured out of a spout and a thermometer that measured temperature by drawing water into a fine capillary tube. Armed with his instruments and his new scientific method, Galileo began deconstructing Aristotlean science.

He set out his ideas about motion in a book called *Two New Sciences*. But – because Aristotle was so revered by Church authorities that any criticism was considered close to heresy – to disguise his own views Galileo wrote a dialogue involving three characters: Salviati, Sagredo and Simplicio. The Aristotelian view was put forward by the simpleminded Simplicio and then criticized by the other characters. Sagredo asserts that, 'I . . . who have made the test, can assure you that a cannon ball weighing one or two hundred pounds, or even more, will not reach the ground by as much as a span ahead of a musket ball weighing only half a pound, provided both are dropped from a height of two hundred cubits.' Although the popular version of this observation is of Galileo dropping objects off the Tower of Pisa to prove his hypothesis, there is, unfortunately, no evidence he actually conducted such an experiment. In fact, in later years he said that the idea first came to him, watching hailstones, noting that large heavy hail hit the ground at the same time as the smaller lighter hail. If heavier objects travelled faster, then the larger hailstones must have either started their descent from a higher point, or the smaller hail began descending earlier. Neither possibility seemed likely, so Galileo devised a simple experiment, measuring the rate balls roll down an slope. He discovered that heavy and light balls rolled at about the same rate. From these observations, he concluded that all objects moving on inclines, or in free-fall, move with a constant acceleration independent of weight.

While Aristotle had thought it necessary to push an object to maintain constant motion, Galileo argued that if there were no resistance then moving bodies would continue to move for ever. 'Imagine any particle projected along a horizontal plane without friction; then we know ... that this particle will move along this same plane with a motion which is uniform and perpetual, provided the plane has no limits.' Galileo employed the same reasoning to introduce his famous principle that all movement is relative. He reasoned that a person waking up in a windowless cabin on a boat moving at a constant speed on a calm sea would have no way of telling whether he was moving at a steady speed in a fixed direction – or was absolutely stationary. The dynamics of movement within the room would be exactly the same. Only by looking out a porthole and observing the relative motion between the ship and any visible land, would the person be able to discover his state of motion. In fact, he could not even be sure that it was the ship rather than the land doing the moving, as it comes to the same thing in Galilean relativity. All he could observe was their relative motion. In modern terminology, the principle is formulated as 'Galilean relativity' which states that the laws of physics are the same in all frames of reference in uniform motion. So movement is relative and needs to be defined with the aid of external references.

In 1642, the year of Galileo's death, Isaac Newton was born in England. He was very premature (his mother said that he could be fitted into a quart pot) and was not expected to survive. Luckily he was stronger than expected and lived to the age of eighty-five. As a young man, Newton's mother removed him from the Grantham grammar school where he had shown little academic promise (described as 'idle' and 'inattentive'), encouraging him to take up farming. His efforts at farming were such a total failure that his uncle persuaded Mrs Newton to send young Isaac to university. In 1661 he went up to Trinity College, Cambridge, paying his way through college by waiting tables and cleaning rooms for the wealthier students. However, in 1665, the plague reached Cambridge and the University was closed. Newton returned home to Lincolnshire for two years and during that time formulated many of his ideas on mathematics, motion and optics.

Newton was a curiously introverted man, reluctant to publish his work; it was often only when a rival published competing work that he went into print himself. So it wasn't until 1687 that he published his great work, the *Philosophiae Naturalis Principia Mathematica* or simply the *Principia*. In it Newton analysed the motion of bodies, including projectiles, pendulums, and the planets. He showed that precisely the same laws that described the movement of a cannonball could also describe the movement of the heavenly bodies. At a single stroke, Newton severed the link between the physical world and superstition – even the heavens were claimed as part of the rational universe. The *Principia* is generally held to be the greatest and most influential scientific book ever written (although I would opt for *The Origin of Species*). Its publication, more than any other event, marked the end of mediaeval dogma and launched science into the Age of Reason.

Newton described three laws of motion. The first is often called the Law of Inertia, and it states 'Any body that is either in constant motion or stationary will remain either in constant motion or stationary unless some force acts upon the body.' If you don't kick a ball, it doesn't move. This Law is just a more precise statement of Galileo's law of motion but it places the concept of *force* centre stage. Constant movement does not need the action of any forces. It is only when movement changes that force is involved.

Velocity describes the rate of movement and is nowadays expressed in units of metres per second. Movement can change in two ways. Firstly bodies can speed up or slow down; and, secondly, the direction of their motion can be altered. Both these changes are described as acceleration. Newton's Second Law states that, 'Acceleration of a body is proportional to the applied force.' When you kick a ball, you apply a force that accelerates the ball. Newton realised that heavier objects require a greater force to move them than lighter objects. You need to kick a football a lot harder than a tennis ball to make it move. *Momentum* is an important concept that describes the product of a body's mass[1] times its velocity. A tennis ball and a football may be travelling at the same velocity, but the momentum of the football will be greater than that of the tennis ball.

Newton's Third Law is often described as the law of action and reaction. If you remember, we described *action* earlier, as the change in movement consequent upon a force acting upon a body – such as kicking a football. But if you have ever kicked a hard ball with bare feet you will surely have discovered that the ball kicks back. This action of the ball on your foot is referred to as a reaction to the force applied by your foot to the ball. Newton's Third Law simply states, 'For every action there is an equal and opposite reaction.' It makes no difference which we consider the action and reaction, and the Third Law is more often stated today in the form, 'When a body A exerts a force on another body B, then body B exerts and equal and opposite force on body A.'

We can now examine our football-kicking action in the framework of Newton's laws. If we consider the ball first, initially it was stationary but it then moved. Its change in motion is acceleration. According to Newton's Second Law, that acceleration requires a force to have acted on the ball. That is what our boot provided. The Third Law says essentially that the force is symmetric; it acts both ways. The reason the ball moved, rather than our body, is that we are more massive than the ball and therefore have a greater inertia (inertia describes the resistance of bodies to changes in motion).

Newton's laws tell us that to understand motion we should look for forces. But what are forces? Newton did provide us with a description of one, gravity, and the famous tale of the apple falling on his head is widely used to illustrate that force's action. But although gravity affects all bodies on Earth, it isn't the force that drives trains, cars, footballs or legs. To discover how these objects move, we must look for other forces.

'THE FORCE THAT THROUGH THE GREEN FUSE DRIVES THE FLOWER'

Magnetism probably corresponds to the image most people have of force, an invisible influence able to move objects without touching them. Newton recognised that gravity was such a force. Electrical force is another. Lightning strikes were, of course, a familiar phenomenon, but it wasn't until 1746 that Benjamin Franklin showed that lightning is a discharge of electricity, demonstrating the existence of electrical force. In 1791, the Italian Luigi Galvani demonstrated that electricity was involved in life's movements: he showed that a dissected frog's leg could be made to move (galvanized) through the application of a charge from a crude electric cell – and thereby inspired Mary Shelley to animate Frankenstein's monster with the 'spark of being'.

The forces of magnetism, electricity (actually two aspects of the same force, electromagnetism) and gravity are all we need to know to understand the movement we see around us. To complete the picture however, there are two additional forces: the weak and the strong nuclear force. These forces are involved in holding together atomic nuclei and are unleashed in nuclear bombs or harnessed by nuclear power.

Most modern physicists believe that there is, in fact, only one fundamental force, and that the four forces are each aspects of it. They search for a theory of everything that unites all the forces. The search started in 1860 when the Scottish physicist James Clerk Maxwell showed that magnetism and electricity were both aspects of electromagnetism. Probably the greatest physicist of the nineteenth century, at the age of three Maxwell was described: 'he has great work with doors, locks, keys etc., and "Show me how it doos" is never out of his mouth. He also investigates the hidden course of streams and bell-wires, the way the water gets from the pond through the wall and a pend or small bridge and down a drain."' The triumph of his scientific career came with the publication of a series of equations that described the behaviour of the electromagnetic force.

One of the most remarkable upheavals in the history of science came when Maxwell found that one of the constants that appeared in his equations was the velocity of light. This was extraordinary since the equations were written to describe the behaviour of electricity and magnetism – nothing to do with light – or so people thought at the time. Maxwell concluded, 'that light consists in the transverse undulations of the same medium which is the cause of electric and magnetic phenomena'.

We now describe light as an undulation (or waving) of the electromagnetic field. You may be familiar with the concept of a magnetic field such as the one that envelops Earth and makes compass needles point to magnetic north. Electrical fields and magnetic fields are two aspects of electromagnetism. The space between charged or magnetic particles is permeated by an electromagnetic field. If you place a positively charged particle in an electromagnetic field, it experiences a force that moves it towards the electronegative pole. So the electromagnetic field is a kind of field of force, or, as science-fiction writers prefer, a force-field. As you can have waves in matter so you can also have waves in the electromagnetic force-field; we see these waves as light.

Maxwell and his contemporaries believed that the electromagnetic force-field was, quite literally, a fluid medium, which they called the ether, that permeated space. Waves would pass through the fluid ether as they pass though fluid water. However, in 1887, two American scientists, Albert Michelson and Edward Morley, attempted to detect the ether by measuring the speed of light in the direction of Earth's motion and at right angles to it. If light was a vibration of ether, then shining the light in the same direction as Earth's motion should make the waves arrive quicker. Michelson and Morley could find no evidence for this effect. There appeared to be no ether. What then was waving to make light waves? The solution to this problem had to await the twentieth century and a remarkable theory from a Swiss patent agent.

The electromagnetic force can be either attractive (between oppositely charged particles) or repulsive (between like-charged particles). Big objects (such as footballs and boots) have an electrically balanced mixture of billions of positively charged protons and negatively charged electrons. The attractive and repulsive forces tend to cancel each other out, leaving zero net electromagnetic force between big objects. However, when the boot gets very close to the ball, the electrons on the boot's surface and the electrons on the ball's surface come close enough to experience a mutually repulsive electromagnetic force – since the surface electrons are both negatively charged. The repulsive force is equally distributed between our boot and the ball (Newton's Third Law) but because we have much greater inertia than the ball, it is the ball that moves. Our boot experiences the repulsive force as recoil, which slows its forward swinging motion. If we weren't wearing boots, this reaction would hurt our toes.

So that is how we move a football without touching it: it is the electromagnetic force that moves the ball. It is relatively easy to see how the action of a force can move a football but how does the electromagnetic force move your car or our bodies? To answer this question, we next need to investigate the *motive power* of fire.

REFLECTIONS ON THE MOTIVE POWER OF FIRE

Kicking is not the only way we can make a football move. If we lit a firework beneath it, the resulting explosion would shift the ball. With a little more ingenuity, we could push the ball along with a toy steam engine. But both these inanimate sources of movement involve fire. To understand how fire moves bodies, we must explore a science whose origins lie in revolutionary France.

One of the great men who rose to power during the French Revolution was Lazare Carnot. He recognized the genius of Napoleon, appointing him to his first command. He was later Minister of War under Napoleon, and organized the armies that conquered much of Europe. Like many of his fellow republicans, Carnot had wide-ranging interests. He loved literature and music, naming his son Sadi after the mediaeval Persian poet, Saadi Musharif ed Din. He was also a brilliant mathematician and engineer, publishing in 1803 a book entitled *Fundamental Principles of Equilibrium and Movement*. In it he discussed how machines convert one form of energy into another.

Carnot must surely have shown his son the gears, pulleys and the other mechanical contrivances of nineteenth-century engineering, because Sadi, in turn, became fascinated by machines. The boy also demonstrated an extreme independence of mind from the age of four. When he and his family were visiting Napoleon, they went on a boating trip. Napoleon was teasing some women by throwing stones into the water to splash them. Sadi fearlessly rushed up to Europe's conqueror, shaking his fist and demanded: 'You beastly First Consul, stop teasing those ladies!' Sadi appeared to have carried this free-spirited defiance into his adult life. He was among the students who took part in the defence of Paris when it was besieged by the Prussians in 1814. He refused to trade on the eminence of his family name, even after Napoleon came to power again during the Hundred Days. Like his father, Sadi was interested in the arts and apparently was an excellent violinist. But his greatest interest was in the motive power of fire.

In 1823, he wrote *Reflections on the Motive Force of Fire*. Carnot's inspiration was the steam engines which had revolutionized industry in

England. He believed that it was England's industrial might, built on steam power, which had ensured victory over France. With his book's publication, he hoped to bring the benefit of steampower to the French. He devised the Carnot cycle, still used today to analyse how engines convert heat into work. He showed that heat cannot pass from a colder object to a warmer and that the efficiency of engines depends on how much heat they can convert into useful work. His ideas laid the foundations for the science of thermodynamics.

Yet hardly anyone bought Carnot's book and it disappeared almost without trace. In 1834, another French scientist Emile Clapeyron, one of the few to have read and been inspired by it, wrote, *Memoir on the Motive Power of Heat*. This too was virtually ignored until, in 1843, it was translated into German and brought Carnot's name to the attention of nineteenth-century physicists. Unfortunately, by then Carnot was dead. He had died in 1831, victim of the second great cholera epidemic to sweep nineteenth-century Europe.

The science Carnot invented, thermodynamics, is an extraordinarily wide-ranging discipline, rivalled only by mathematics in its pervasive influence over phenomena as diverse as mechanics, chemistry, biology, ecology, geology, weather, history of the universe, origin of time and the dynamics of black holes. Thermodynamics is what makes cars and steam engines move and fireworks explode. It can – like a leg – move footballs. The key question for us is whether thermodynamics explains biological motion.

At its heart, thermodynamics is extremely simple. In fact, it is the same phenomenon going on, over and over again in a million different guises. Its most straightforward manifestation can be seen if we consider a box with a partition down the middle, forming two smaller chambers. (FIGURE 6.1A) The left-hand chamber is filled with a gas; the right-hand chamber is empty (a vacuum). If we imagine the gas as air and the box as a cubic metre then the left-hand chamber will contain trillions of gas molecules (roughly 10^{27} molecules) all zipping around, bumping into each other and the container walls. The motion of individual gas molecules is entirely random or incoherent; some will move to the left, some to the right or up or down. Because there are trillions of gas molecules, their individual motions will cancel each other out, so the bulk volume of gas does not move (also it can't because it is inside the box). Whenever a molecule bumps into the walls, it will bounce back. But the collision will not be entirely elastic: some energy will be transferred to the walls. We feel this energy as heat. If we light a Bunsen flame in the left-hand chamber then the air molecules inside the box will absorb some of the flame's energy. This energy will make the air molecules move around faster, with more kinetic energy. Those gas molecules that bump into the box sides will transfer more energy to the walls and the box will get hotter.

What happens to our box of air if we remove the partition? The gas molecules in the left-hand chamber continue to move with exactly the same incoherent motion as they did before. They will zip about, moving in all directions inside the box, bumping into each other and into the sides of the container. However, the right-hand partition they used to bump into, is no longer there. Those molecules that previously would have hit the partition and bounced back, are now free to move into the right-hand end. There will therefore be a movement of gas from the left-hand area into the right-hand (FIGURE 6.1B), until the entire box is evenly filled with gas (FIGURE 6.1C).

It is important to realise that the motion left → right is only apparent at the level of the gas's bulk volume. If we examined an individual gas molecule just after the partition was removed, we would be as likely to find it moving to the left, to the right, up or down, or indeed in any direction. The reason the bulk volume of gas moves is simply that there is more empty space available to the right for the gas molecules to move into. There is *no directional force* that causes the motion of molecules from left to right.

The apparent motion of the gas can also be considered in terms of changing probability. Whereas, before we removed the partition, there was a zero probability of finding a gas molecule in the right-hand chamber, that probability increases steadily once the partition is removed. The motion of individual molecules remains entirely random, but their aimless motion leads to a steadily increasing probability of finding a gas molecule in the right-hand side. This involvement of probability introduces us to the extremely important concept of *entropy* and the all-pervasive Second Law of thermodynamics. The Second Law states that natural processes are accompanied by an increase in entropy of the universe.

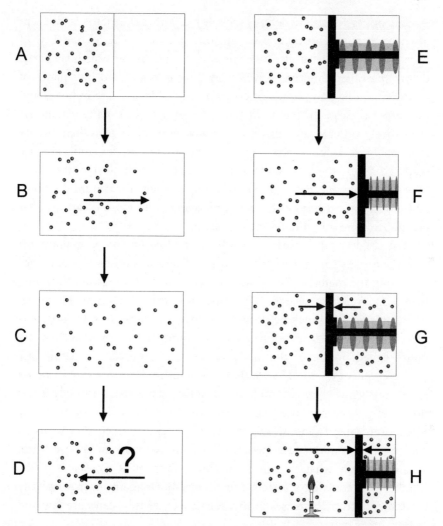

FIG 6.1   The action of thermodynamics. All thermodynamic-driven events involve a redistribution of matter and energy. In A, gas molecules are enclosed in the left-hand chamber of a box. If the partition is removed then random molecular motion will fill the entire box (A→B→C). There remains a small but finite probability that random fluctuations will cause all the gas molecules to spontaneously re-enter the volume of the left-hand chamber (→D). In E→F, the kinetic energy of gas molecules in the left chamber is captured to do work (compress a piston). In G→H, heat is used to increase the kinetic energy of gas molecules in the left chamber and so form a heat engine.

Entropy is difficult to define precisely but comes closest to our concept of chaos. It is much easier to see entropy in action. The Second Law of thermodynamics reveals itself whenever my son Ollie has a birthday party. Before the other children arrive, everything is neatly laid out: plates of sandwiches, cakes and bowls of crisps on the table, toys are in boxes and the carpet is clean. In thermodynamic terms, we can say that the house is relatively ordered and thereby in a low entropy state. When twenty children leave a few hours later, the sandwiches, crisps and cake are scattered across the table and carpet; toys are strewn all over the floor; Ollie's bedroom has been gutted and all its contents thrown on the floor; and we all have a headache. In thermodynamic terms, we can say that our house is now in a high entropy state.

The children who dispersed this food and toys had no particular wish to make a mess (or at least most didn't). There was no directional force dispersing the material. The children merely picked things up, dropping them wherever they pleased. It so happens that of all the possible places where toys or food may be dropped, only a very few are tidy (ordered) places. In itself there is nothing special about the untidy state that my house ended up in. In thermodynamics terms, we cannot define any characteristic of an untidy state that distinguishes it from a tidy (ordered state). But while there are millions of untidy states there are only a few tidy states. When my house started in one of the tidy states and the entry of children caused random motion, then it was very likely that it would end up in an untidy state, because there are so many to choose from. This is basically what entropy is all about. Entropy is a measure of the number of states that a system can occupy but remain macroscopically the same. Entropy increases in my house because tidy states are few and untidy states many.

For our box of gas, we can consider entropy to be a measure of the number of possible arrangements of gas molecules. When the gas was confined to the left chamber, there were perhaps a billion (the actual figure would, of course, be very much larger) possible ways the gas molecules could have been arranged. Removing the partition vastly increased the number of possible arrangements – perhaps to a trillion. Of all the trillion ways now possible to arrange the molecules, only a billion correspond to the original arrangement with all the molecules confined to the left-hand chamber. Random incoherent movement thereby disperses gas molecules into the vastly more numerous states filling the entire box. Entropy thus increases.

The Second Law of thermodynamics states that everything that happens in the universe is accompanied by an increase in entropy. This may appear somewhat sweeping, yet, as far as we know, it is true. It implies that all physical change involves the same dispersion that occurred at my son's party. Of course, the dispersion will seldom involve sweet-papers or cake, it need not even necessarily involve matter, but something is being dispersed whenever anything happens in the universe. The all-pervading influence of the Second Law has profound significance for our understanding of the universe, how we came to be here, the direction of time, and the universe's ultimate fate. It is the reason thermodynamics is involved in so many diverse physical phenomena. Indeed the Second Law demands categorically that it must be involved in all physical phenomena including, of course, life.

To see how entropy is involved in heat engines, imagine our box again but now with a piston (FIGURE 6.1E) replacing the partition wall. It is now the piston the rightward-moving molecules are bumping into, rather than a wall. Newtonian mechanics tells us that whenever each molecule bumps into the piston, it will exert a force on the piston equal to its mass times its acceleration. The combined actions of billions of incoherently moving molecules bumping into the piston will force the piston to move to the right. We could harness the piston's kinetic energy to do some work such as drive a car or move a football; or we could store that energy in the potential energy of a spring (as shown in the figure). Again, it is not a directional force that causes the directed motion of the piston from the left to the right: it is driven instead, by the random motion of billions of particles.

But, although we have moved a piston, we do not yet have a heat engine. To convert our box into a practical heat engine, all we need to do is imagine the two chambers separated by a piston as both filled with similar numbers of air molecules (FIGURE 6.1G). Now the force exerted by molecules hitting the piston whilst moving to the right (in the left chamber) is exactly equalled by the force of molecules hitting the piston whilst moving to the right (in the right chamber). The incoherent motion of the air molecules can no longer be harnessed to move anything. However, if we now add a Bunsen flame to speed up the molecules' motion in the left chamber, their greater kinetic energy will exert a greater force on the piston, causing it to move to the right. This can be harnessed to move a ball, power a train or compress a spring. This is the principle of the heat engine. Once again, the piston's motion is not caused by any directional force, but by the incoherent motion of billions of particles towards increasing entropy. It is the structure of the box, piston, chambers and the positioning of the flame that gives this motion its direction.

All heat engines work on the same basic principle. They convert the random motion of billions of molecules into the directed motion of the moving parts of the engine. But not all heat engines are steam engines. We could (with care) use a petrol-engine powered car to push our football along. Nevertheless, chemical heat engines, like car engines, are also powered by entropy. The difference is that in chemical engines, energy, as well as molecules, is dispersed. Car engines work by burning fuel, petrol (gasoline) or diesel. Both fuel and air are injected or blown into the piston chamber and this mixture is ignited. (Chapter Five briefly described burning.) High-energy electrons in molecules of fuel move to low-energy orbitals in oxygen. In the process, the energy of the fuel electrons becomes dispersed amongst all of the molecules (oxygen and fuel) in the chamber. The energy that before ignition was locked up within a (comparatively) small number of molecules (in the fuel) is now dispersed amongst a much larger number (the products of combustion). Entropy increases.

It is important to realize the Second Law does not forbid processes that decrease entropy but states that such processes must be balanced by a coupled increase in entropy. I can tidy up my house after my son's party and thereby decrease its entropy but in so doing I will be hydrolysing ATP in my muscles and burning glucose. These chemical reactions are accompanied by large increases in entropy (energy becomes more dispersed); so on balance, the Second Law is not violated. Many entropy-decreasing chemical reactions can be similarly coupled to entropy-increasing chemical reactions. The chemical wave-generating Belousov-Zhabotinsky reaction described in Chapter Four is one such. Chemical reactions taking place within the dish are dispersing energy; the increasing entropy associated with the dispersal is coupled to the entropy-decreasing reactions generating the waves of colour.

The structures generated by entropy-coupled self-organization, such as the Belousov-Zhabotinsky reaction, were called dissipative structures by the Nobel prize-winning Belgo-Russian chemist Ilya Prigogine. This term comes from their tendency to dissipate entropy. Dissipative structures are widespread in nature. The pattern of convection flow when liquids are heated from below is a dissipative structure. You must surely have seen (you might even own) one of those lava lamps filled with brightly coloured immisible liquids. When the lamp is turned on, the liquids are heated and the convection flow generates swirling blobs of colour. Convection currents form because ordered convection flow transports heat more efficiently from the cold to the hot surface than disordered flow. The heat transport increases entropy for the entire body of liquid plus its surroundings; this entropy increase more than balances the decrease associated with the ordered flow. Weather patterns, such as the Azores anticyclones, are examples of dissipative structures. Heat engines can also be viewed as a kind of engineered dissipative structure.

Dissipative structures depend on a continuous flow of energy into the system. The lava lamp blobs are maintained by heat energy; both heat energy and the kinetic energy of the wind maintain weather patterns; the Belousov-Zhabotinsky reaction is maintained by chemical energy. Without a flow of energy into each system, the ordered structures dissipate. Turn the light (and heat) off and the blobs fall to the base of the lamp. Extinguish the fire and the steam engine stops. This dependence on a flow of energy into – and out of – the system is very reminiscent of the dependence of biological systems on an energy source. Dissipative structures are also akin to the phenomenon of self-organization which complexity theorists such as Stuart Kauffman have proposed as the basis of life. Ilya Prigogine (amongst others) claims that life itself is a dissipative structure.

ARE WE HEAT ENGINES?

As far as we know, the Second Law of thermodynamics is never violated. Every reaction performed by chemists is driven by entropy increase. It is therefore hardly surprising that thermodynamics has been brought to bear on life. The most striking feature of living systems, from a thermo-dynamical point of view, is their startling degree of order. As a creature grows from a single cell to a complex organism, its entropy, already low, decreases to an astonishing extent. So what happens? Does life break the Second Law of thermodynamics?

The answer is no. The low entropy state of our bodies is coupled to entropy increases elsewhere. The chemical reactions that take place when our food is converted to waste products generate huge increases in entropy which more than balance our body's low entropy state. The Second Law is not violated. In this sense we are dissipative structures that dissipate entropy. But is the order of living cells dependent upon the same dynamics that generate convection flows, chemical waves, anticyclones or drives steam engines – the random motion of billions of particles?

Many scientists believe that life is exactly this, an elaborate heat engine. For instance, the chemist P. W. Atkins writes in his (very readable) account of thermodynamics, *The Second Law*, that 'The process of biology and chemistry, rich and extraordinary as they may seem ... are no different in principle from cooling.' And later, in a rather melancholy tone, 'We are the children of chaos, and the deep structure of change is decay. At root, there is only corruption, and the unstemmable tide of chaos.' Or from Stuart Kauffman, 'cells are nonequilibrium dissipative structures.'

Yet I hope you have spotted an essential, *crucial* difference between thermodynamic-driven dissipative structures and the directed actions of life. Steam engines, car engines, the Belousov-Zhabotinsky and the Azores anticyclone *are driven by the random motion of billions of particles* – they represent order from disorder. While heat or chemical engines generate motion harnessing the chaotic movement of billions of particles, enzymes like myosin generate motion by *directing* the motion of single protons, electrons or ions. Mitochondria work by directing the motion of indi-vidual protons across the mitochondrial inner membrane. The enzyme LDH works by directing the motion of single electrons within its active site. Muscles contract by directing the motion of protons within the myosin hinge region. The motion of single particles in the DNA molecule causes mutations. Thermodynamics is a science of big numbers. It is about the bulk properties of matter and the order it generates is only an average kind of order. At the level of fundamental particles, everything is chaotic. In contrast, life is a phenomenon of small numbers and displays order, right down to the level of fundamental particles. As Schrödinger stated in 1944, life is 'order from order'.

Entropy and thermodynamics are certainly involved in life just as they are involved in every other phenomenon in the universe. Gravitational collapse, nuclear dynamics and planetary motion all obey the Second Law. Yet thermodynamics does not *explain* radioactive decay, black holes or the rings of Saturn. Similarly, thermodynamics does not explain life. Living cells are no mere heat engines.

But if thermodynamics isn't driving the motion that goes on inside our cells, then what is?

THE DEMON IN THE BOX

Life cannot escape the rule of the Second Law of thermodynamics but might be able to side-step it. Consider again our box of air after the partition has been removed and the molecules have become evenly distributed FIGURE 6.1C. Can all the gas molecules spontaneously return to the left-hand area FIGURE 6.1D? The answer is, in fact, yes. But remember that there are 10^{27} molecules of air in the box. The probability that the random incoherent motion of each of the 10^{27} molecules could end up with their finding their way into the left end of the box is very, very, very, small – *practically* zero. This is fortunate because it is more or less the same probability that all the molecules of air in your room would spontaneously rush out of your window, leaving you in an uncomfortable airless vacuum. Such bizarre occurrences are possible, but not at all likely.

However, if, instead of trillions of molecules in the box, we have only ten or a hundred (more like the actual number drawn), then the probability of their all finding their way into the left-hand end of the box is not that small. Random fluctuations would, every now and again, result in all molecules being located in the left-hand chamber. This would lead to a transitory decrease in entropy, reversed a few moments later when the molecules bounced back into the right-hand chamber. But, consider what would happen if a microscopic demon were also inside the box, who could slam down the partition whenever he noticed this transitory deviation from random distribution. The demon could capture all the molecules in the area of the left-hand chamber and effect a permanent decrease in entropy, in apparent contradiction of the Second Law.

This demon has a name. He is called Maxwell's demon after James Clerk Maxwell, (he of the electromagnetism equations). Maxwell pondered whether the Second Law could ever be violated by the action of such a demon. A great deal of ink has been spilt on this question and the consensus answer is no. The problem for the demon is that he has to gain information about the molecule's positions in the box by some kind of measurement. This process of gaining information (perhaps by shining lights on the particles to discover their position) must then cause an increase in entropy which compensates for any decrease in entropy he achieves.

Nevertheless, I believe Maxwell's demon has profound significance for life. I will be proposing that, like Maxwell's demon, living cells are able to capture low entropy states. But, like Maxwell's demon, living cells must also gain information about the motion of their fundamental particles by some kind of measurement process. How does a cell *measure* the movement of a proton or an electron? How does anything measure fundamental particles? To answer these questions we must finally turn to that strangest of sciences, quantum mechanics.

7

What is Quantum Mechanics?

At the end of the nineteenth century, Lord Kelvin, one of that century's greatest physicists, claimed that the basic outline of physical theory was more or less complete. The giants of eighteenth- and nineteenth-century science had discovered the equations of motion and laws of change that could be used to predict a cannon ball's trajectory, the motions of the planets or determine how much heat was required to power a steam train. Nature's grand plan had been laid bare. Henceforth physicists would preoccupy themselves with the details. Kelvin noted however that there remained 'two small clouds' on the horizon – two problems that had yet to be solved. He did not doubt these would soon be dispersed by the relentless progress of mechanistic physics. But it is perhaps an ironic tribute to Kelvin's genius, that he had correctly identified the precise problems – those 'two small clouds' – whose solution would lead to the startling deconstruction of the entire edifice of nineteenth century physics.

The first of Kelvin's 'clouds' was the experiments of Michelson and Morley, mentioned briefly in the last chapter, which could find no evidence for an ether through which electromagnetic waves might pass. Einstein's special and general theories of relativity proposed solutions would revolutionize concepts of space and time.

The second 'cloud' was seemingly much more mundane: it was classical theory's failure to account for the spectrum of radiation emitted from a 'black body'. A black body is an entity somewhat inappropriately named since it need not be black. It is merely an object that absorbs all incident radiation and so appears black in reflected light. However, if you heat it, it will glow – the sun is, in fact, a pretty good approximation of a black body. Indeed, this argument about black body radiation can be broadened to include the radiation emitted from many other hot objects, such as metal. Heat a lump of metal and it glows – *red-hot*. The increasing thermal energy causes the atoms within the metal and their associated electrons to vibrate. Vibrating electrons are accelerating charges. One of Maxwell's equations predicted that accelerating charges radiate electromagnetic energy – the atoms within hot metal radiate light. The frequency of the light corresponds to that of the molecular vibrations. Hot metal glows red-hot and then white-hot as increasing thermal energy causes the atoms to vibrate faster and faster, emitting radiation at higher and higher frequencies. Classical physics predicted that hot black bodies would emit immense quantities of very high frequency ultraviolet radiation – which was called the ultraviolet catastrophe. But no such catastrophe had been seen: instead, most radiation emitted was in the visible part of the spectrum.

MAX PLANCK AND THE END OF THE CLASSICAL WORLD

In 1874, when Max Planck was considering a career in science at Munich University, he discussed the prospects of research in physics with the physics professor, Philipp von Jolly. Von Jolly's advice was bleak: to choose another discipline, since physics was essentially complete with little prospect of further developments. Planck was not dissuaded. He later described his reasons: 'The outside world is something independent from man, something absolute, and the quest for the laws which apply to this absolute appeared to me as the most sublime scientific pursuit in life.' Ironically, Planck's own later work would seriously undermine this belief in an 'outside world . . . independent from man'.

In 1900, Planck came up with an equation that predicted the correct spectrum of black body radiation. However, its implications were so startling that even he was reluctant to accept them. Previous attempts to model the black body radiation assumed that light energy was released continuously from hot bodies. Just as increasing the speed of a motor boat's propeller causes a steady increase in water turbulence so the increasingly accelerating electrons in a hot body were thought to cause a continuous increase in an electromagnetic field's turbulence. However, Planck's equation worked only if the radiation from hot bodies was released in minute packets of discrete size. Planck termed these packets of light energy, *quanta*. Each *quantum* of light carried one discrete energy packet (E) of magnitude equal to the frequency (v) of the radiation multiplied by a new fundamental constant (h), thereafter termed Planck's Constant: $E = hv$. This interpretation seemed to contradict the wave theory of light derived from Maxwell's equations. How could waves come in *packets*?

Planck was a conservative physicist brought up in the nineteenth-century tradition, and was reluctant to relinquish the wave theory of light. He believed instead that the problem was down to the vibrating molecules and that it was these that had the odd property of emitting light in discrete quanta. But this only highlighted an even more fundamental problem. If vibrating molecules emitted light in discrete energy packets that implied that their vibrational energy was similarly quantized. This was an entirely new concept to classical physics, where it had always been assumed that energy could take a continuous range of values.

Consider a molecule shaped like a dumbbell. Such a molecule can wiggle about in different ways, lengthening, shortening and spinning. These movements are termed the *degrees of freedom* of the molecule. Classical theory predicted that the series of radiation frequencies, the spectrum emitted by such a molecule, corresponds to the frequency of these wiggles. Molecules were expected to spin at a continuous range of speeds, from slow to fast, depending on how much energy was applied, like a propeller. But Planck's equation implied that molecules could only spin at certain discrete speeds. There was a minimum spinning speed (the ground state) and then a series of discrete multiples of this minimum speed. As more energy was applied, the molecule would *jump* instantaneously from one speed of spinning to the next state up, generating the discrete spectrum of emitted light. These *quantum jumps* were a unique feature of quantum mechanics unprecedented in classical physics. What happens to the *inbetween states*? How does a molecule instantly switch from one speed of spinning, one *quantum state*, to another? Certainly, no propeller can do that trick.

Nevertheless, Planck's introduction of the concept of quantized energy and quantum jumps proved extraordinarily fruitful. In 1910, New Zealand physicist Ernest Rutherford was working at Manchester University with Hans Geiger (of the counter fame) and Ernest Marsden investigating the structure of atoms by bombarding high-energy alpha particles (helium nuclei consisting of two protons and two neutrons) at thin sheets of gold foil. Although most particles went straight through the metal (since – like all matter – it was mostly empty space) a few were deflected at sharp angles, indicating that they had collided with a very dense core among the atoms of the metal. Rutherford proposed that atoms were mostly empty space with a small dense positively-charged nucleus surrounded by a swarm of negatively-charged orbiting electrons. The problem with their model was that positive and negative electrical charges attract. Electrons orbiting the nucleus should (since they are accelerating charges) lose energy by emitting electromagnetic radiation, spiralling down into the positively-charged nucleus. What was keeping the electrons up? In 1913, the great Danish physicist and chief architect of quantum theory, Niels Bohr, proposed a radical solution. Electrons bound to nuclei were no longer allowed to take on a continuous range of energies but were, like light, quantized at discrete energy levels. Electrons could not 'spiral down into the nucleus', since this would involve their taking on a continuous range of (decreasing) energies. Only quantum jumps from one energy level to another were allowed, and each jump must be associated with the absorption or emission of a photon of electromagnetic energy (with the frequency given by Planck's equation).

The model worked and could be used to predict accurately the spectrum of light emitted by hydrogen atoms. But *why* did it work? Why was the energy of electrons bound to atoms quantized whilst free electrons were allowed to take on any value? What was stopping the electrons from taking a quantum jump to zero energy by jumping into the nucleus? Bohr had constructed a model that fitted the data, but it left many questions unanswered.

So Planck's quantization rule worked for accounting for discrete energy levels in atoms and molecules, but nobody really knew why. Albert Einstein was the first to take the equation at face value, claiming that light itself came in quantum packets. In 1904 he proposed a radical solution to another problem in classical physics, the *photoelectric effect*. Classical theory predicted that if light of sufficient intensity was shone onto a metal's surface, then electrons should be knocked out of the metal by the energy of the light waves. The greater the intensity of light, the greater the energy that should be imparted to the departing electrons. In the same way, waves on the seashore will throw pebbles onto the beach. The bigger, more powerful waves are capable of tossing pebbles faster and further up the beach. However, increasing the intensity of light increased only the number of electrons emitted from the surface of metals, not their energy.

Einstein's solution was to propose that light itself came in discrete quanta or *photons*. Instead of a wave, Einstein envisaged light as a stream of photons. In the photoelectric effect, each photon arriving at the metal's surface would deliver a discrete quantum of energy – equal to Planck's Constant *times* the light frequency, $E = h\nu$. When an electron absorbs a photon of light it gains a quantum of light energy, and carries this amount of energy when ejected from the metal. The light's intensity (number of photons) may increase the number of electrons ejected, but not their energy – that is determined solely by their photon energy and hence the frequency of light. He published his paper on the photoelectric effect in 1905. In the same issue of the journal *Annalen de Physik*, he published two further papers: the first proved that atoms were real and the second was a description of his theory of special relativity. It was a busy year for Einstein, but it was for his work on the photoelectric effect that he received his Nobel prize in 1921.

With Einstein's photon, we go back to an earlier particle theory of light. Newton was convinced that light was made up of particles, yet by the time of Maxwell's electromagnetic equations it was generally accepted that light was a wave. This was due to a very simple experiment performed by Thomas Young in 1802. Young was a remarkable man who could read by the age of two.[1] Trained as a physician and ophthalmologist, he contributed greatly to our understanding of the eye. But the experiment for which he is justly renowned involved shining light through holes. It is a remarkable fact that shining a light through a hole can reveal truly deep mysteries about the world we live in.

MAKING LIGHT OF A HOLE

To predict the future trajectory of any moving body, whether a truck, a photon, or electron, all we need to know is its position at any point in time, together with its momentum and the direction and magnitude of any forces (e.g. those applied by brakes) acting upon it. Insert these values into the equations of motion and that body's future trajectory can be calculated. Scientists worried about asteroids approaching Earth are busily engaged in measuring position and momentum for these rocks in an effort to discover those whose trajectory might put them on a collision course with Earth.

Imagine shining a strong beam of light, say, from a torch or laser pointer, through a one-millimetre-square hole in a screen, to illuminate a white wall. The light beam forms an image of the hole on the wall. Our aim is to determine both position and momentum for photons at the moment they pass through the screen, assuming that no forces are acting upon the photons.

The first thing we can be sure of is that any photons of light that strike the wall must have passed through the hole (to ensure this we could extend our screen as far as we like, in all directions). This makes our position measurement for the photons passing through the screen very simple – each photon must be confined to the one-square-millimetre area of the hole. When we perform the experiment, we see a sharp image of the square hole projected onto the wall. A straight line can be drawn from source, to hole, to image, confirming the belief that light travels in straight lines. The image is the same size as the hole. If the photons that passed through the hole possessed momentum either veering left or right or in an upward or downward direction, the light would have spread out as it emerged from the hole. We would see a larger *diffuse* image. The sharpness and fidelity of size of the image demonstrates that the photons possessed zero momentum in the horizontal or vertical plane – only forward-directed momentum. We have thus made a position measurement to localize photons to within one square millimetre and determined their momentum (in either the horizontal or vertical direction) to be zero.

But are we really measuring the position and momentum of photons, the particles of light? To prove real photons are involved we must reduce the light intensity to a level at which only single photons would be expected to pass in single file through the hole. Our problem is then to register where single photons strike the wall. Our eyes, though highly sensitive, are not able to detect light from a single photon. Frogs have much more sensitive eyes, apparently capable of seeing single photons, but make inscrutable laboratory assistants. We will use instead a photographic plate. When a photon strikes photographic film it causes the blackening of a silver grain within the photographic emulsion to form a black dot (it actually takes five to ten photons to expose a grain of silver, a fact we will ignore for greater clarity). To see the tracks of protons, we will repeat our experiment using a very low-intensity light beam and record photon landings on the photographic plate. As before, each photon that alights on the photographic plate must have passed through the one-millimetre hole in the screen, so our position measurement is unchanged. To measure momentum, we examine the position of the individual silver dots on the photographic plate. We discover that they are formed in precisely the same area as the previous bright image of the square hole. A straight line can be drawn from source through hole to black dot confirming that individual photons possess zero momentum in the horizontal and vertical planes. We have therefore successfully performed both a position and momentum measurement on the photons of light.

However, a position measurement of 'within a square millimetre' is hardly impressive on the scale of photons. Our aim will be to reduce the uncertainty of our position measurement. Let us simplify matters by henceforth ignoring uncertainty in the vertical direction, concentrating only on reducing uncertainty in the horizontal (left-right) plane. To reduce this uncertainty, all we need to do is reduce the width of the hole. With each reduction in width, we confine the light to a narrower space within the screen, thereby reducing the uncertainty of our horizontal position measurement. So we pursue this course of action by shining our strong beam of light through a series of narrow slits. At a width of half a millimetre, we again form a sharp image on the wall, this time of a strip half a millimetre across, indicating that the momentum of the light remains zero in the horizontal plane. Yet our position uncertainty has been halved. However, as we continue to reduce the width to below a tenth of a millimetre, the image of the slit refuses to similarly reduce. In fact, as the slit becomes smaller, the image – although fainter – gets bigger![2]

What we are seeing is the well known phenomenon of *diffraction*, whereby light beams *bend* when they pass through a narrow aperture. The observation of the light diffraction was a nail in the coffin of Newton's corpuscular theory of light. Even in the eighteenth century, it was well established that waves could be diffracted by their passage through narrow apertures. Waves – like sound waves – *bend* around corners. The finding that light was also diffracted seemed to clinch the case for a wave theory for light.

But how does diffraction affect our measurements of position and momentum? Our position measurement remains secure, since we have no reason to doubt that the light still had to pass through the narrow slit. However, the zero momentum we deduced for the earlier experiment was based on the sharp image the same size as the slit. Now with a very narrow slit but wide image, we can no longer conclude that photons have zero momentum in the horizontal plane. Some must have been deflected to the left or right as they passed through the slit. Such photons possess momentum (for momentum includes direction) in the horizontal plane. We can no longer say that all photons passing through our slit have zero momentum. Their momentum is now uncertain to a degree determined by the width of the diffracted image.

We can continue this experiment with increasingly narrow slits to gain a tighter and tighter fix on our photon's position. However, with every reduction in the width, the image becomes more diffuse. It seems any reduction in uncertainty in position has to be balanced by an increase in uncertainty of momentum. We can reduce the uncertainty in momentum but only by expanding the width of the slit and thereby increasing the uncertainty of position.

This, of course, remains fully in accord with the wave theory of light – diffraction is greater for narrower apertures. But we have been trying to measure position and momentum for the corpuscles of light, photons. Can the particles of light show diffraction at scales of 0.1 millimetre, surely massive in comparison to their size? Maybe they don't? Water waves are diffracted when they emerge from a narrow harbour mouth but you would not expect the same behaviour from individual water molecules fired by some kind of molecular water pistol through the same

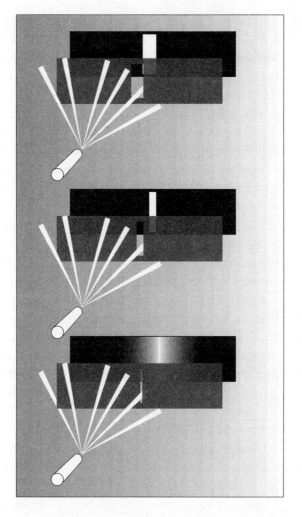

FIG 7.1   Single slit diffraction of light. When a beam of light is
shone through a wide slit (top), a sharp image is formed on the
screen. As the widths of the slit is reduced (middle and bottom),
the image of the slit on the screen becomes more diffuse.

opening. Water waves are a manifestation of the collective dynamics of
billions of water molecules. Perhaps light similarly consists of a dense
swarm of photons and it is their collective behaviour, rather than the
individual photons, that is wave-like.

To test this hypothesis, we can reduce the light intensity again, so
that only single photons pass one at a time through the slit. After the

experiment, when the positions of the dots representing photon landings on the photographic plate, are measured – they fall anywhere within the large image *of the diffracted light beam*. We can no longer draw a straight line between the dots, the hole and the source. Photons are diffracted just like beams of light, with exactly the same implications for measurements of position and momentum.

When light was considered as a wave, diffraction emerged naturally as a consequence of wave dynamics. But the photoelectric effect, the black body radiation problem and many other experiments clearly tell us that light consists of particles. The photographic image used to record our photons is composed of millions of tiny black dots of silver. These dots appear to record collisions between *particles* of light and grains of silver bromide in the photographic emulsion – how else do you produce dots? One possible resolution is to consider that photons are actually *big* particles. If their dimensions were close to those of the slit, then we might expect them to be deflected as they bounced and squeezed through the aperture. Photons do not really possess a definable size, since they can never be confined to have their dimensions measured, but atomic nuclei that absorb or emit single photons are less than a billionth of the diameter of our 0.1 millimetre slit. Diffraction of a particle the size of a nucleus passing through a 0.1 millimetre aperture is equivalent to a bullet being deflected by its passage through a gap a thousand miles wide. How does a photon even *know* that it is passing through such a wide gap? The diffraction effect makes sense only if we consider that a diffuse wave is travelling through the slit. Thus we are left with a puzzle. Light appears to be emitted as a stream of particles; it arrives at a photographic plate in the form of particles; but in its inbetween state, it travels as a wave. Light appears to be schizophrenic.

UNCERTAIN OR UNREAL?

The great German physicist Werner Heisenberg pondered the limitations this kind of experiment imposed on the amount of information we are able to extract from the world around us. In 1927 he presented his famous *Uncertainty Principle*. One formulation states that whenever we make a measurement, then the product of the uncertainty in position and the uncertainty in momentum (uncertainty in position *times* the uncertainty in momentum) must always be equal to or greater than the number of Planck's Constant (*h*) divided by 2π, a value known as $\hbar$ (h-bar). This relationship implies that the units of $\hbar$ are those of distance (of any position measurement) *times* momentum. In physics the product of distance and momentum is, as you may remember from earlier chapters, known as *action*. Therefore $\hbar$ defines the minimum possible uncertainty in action, *the quantum of action*.

Planck's Constant (0.00000000000000000000000000000000066) is such a small number that we only ever need worry about the limitations imposed by its value when we wish to measure very tiny bodies with similarly small values for their actions. With large objects, uncertainty at this minute level is negligible.

However Heisenberg's uncertainty principle is most important when considering the behaviour of the microscopic constituents of matter. For particles with very small mass (such as electrons) and correspondingly small momenta (momentum = mass × velocity) the value of $\hbar$ places very severe restrictions on the accuracy by which we can measure any action. When we passed photons through a wide aperture without attempting to gain accurate position information, the uncertainty principle allowed us to achieve high accuracy for our momentum measurement. The product of the small uncertainty in momentum *times* the large uncertainty in position remained well above the value of $\hbar$. However, as we squeezed our position measurement, the uncertainty principle forced the photons to adopt a wider uncertainty in their momentum to ensure that the combined uncertainty in position and momentum of our measurement remained greater than $\hbar$. It is this measurement-induced increase in the uncertainty of momentum that caused the deflection of photons of light as they passed through the narrow aperture. This uncertainty caused the *wave-like* diffraction of photons of light. The uncertainty principle and the wave properties of light are actually two expressions of the same fundamental reality.

The restrictions imposed by the uncertainty principle apply to all particles, not only to photons. Experiments such as the one described have been repeated with electrons, neutrons and protons and even whole atoms; and they all show the same diffraction effects. The radius of a neutron can be measured indirectly and is about 0.0000000000001 milli-metre. Yet neutrons are diffracted by passage through a hole the size of a postage stamp.

The uncertainty principle restricts our ability to measure not only position and momentum simultaneously, but also other complementary properties. Simultaneous measurement of energy and time are similarly restricted by the uncertainty principle. Measurement of the direction of polarization of light or the magnitude of spin (angular momentum) for electrons or protons, in two complementary planes, is also restricted. Heisenberg's colleague and mentor, Niels Bohr, discovered that all measurements affect two related but complementary properties (such as position and momentum), so measurement of one causes uncertainty of the other: the principle of complementarity.

It is often stated that Heisenberg's uncertainty principle is caused by our inability to measure a body without disturbing it. Consider measuring the position of a stationary billiard ball (with zero momentum) hidden beneath an open-sided box. Our measurement might consist of firing other *measurement* billiard balls, say *m*balls, through the box and noting the direction by which the *m*balls are deflected by their passage through. If we then plot the line of our incident *m*balls into the unknown area (of the box) and similarly project the line for any deflected *m*balls that emerge from it, then the lines cross at the point of collision. We have thereby measured the position of our billiard ball. However, our *m*balls carry a momentum that would have been transferred to our target ball during the collision – it would have been moved. The billiard ball's momentum would no longer be zero. This result would seem to be in accordance with the uncertainty principle's contention that we cannot simultaneously measure both position and momentum for our billiard ball.

However we can improve matters. Instead of *m*balls, we could use ping-pong balls. Since these are much lighter they would carry less momentum, disturbing our target billiard ball far less. Ping-pong balls would allow the same accurate measurement of position but with neglig-ible disturbance to our target's momentum. But why stop with ping-pong balls? We could fire much lighter projectiles such as single atoms, or electrons or photons and measure their deflection by our target. This is of course exactly what we do when we see an object. Then, the photons are usually fired not by us but by the sun or a lamp, and it is our eyes acting as measuring devices. Photons carry very small packets or *quanta* of momentum; so when using them for measurement, we can be confident that the collision between the photons and any heavy object will disturb the object by only a very small amount. The amount of disturbance is relatively easy to estimate since the photon's momentum is given by Planck's equation: the constant $\hbar$ divided by the wavelength of the photon. This level of disturbance, by a single photon, represents the minimum level of interaction between any body and a measuring device. It is no accident that Planck's constant turns up in the uncertainty principle. All measurements must involve some exchange of energy and the uncertainty principle quantifies the minimum disturbance possible when such an exchange takes place.

For a billiard ball, the small amount of momentum uncertainty imparted by collision with a photon is tiny – far smaller than we could detect. However, for an electron (such as those involved in enzyme action), $\hbar$ is very significant. Any photon colliding with an electron will impart a significant quantum of momentum to the electron. Any accurate measurement of position for an electron will change its momentum and future trajectory. If we were to measure the position of a free electron to an accuracy of say, the width of an atom, the momentum uncertainty imparted by our measurement would instantly fling the electron away at such speed that in another second it could be anywhere within an area the size of a small town.

The uncertainty principle is inescapable. Its effect may be negligible for heavy objects but it is always there in the background generating tiny amounts of uncertainty. For microscopic particles its effect can never be ignored. But Heisenberg's uncertainty principle is not merely about our inability to measure a system without disturbing it. The uncertainty principle requires a far more profound shift in our view of reality than any recognition of the limitations of measuring devices. I may be uncertain about many things: the size of my next telephone bill, the capital of Azerbaijan, how many pints of milk remain in my fridge. I am confident however that each of these properties of the exterior world has an objective reality, whether or not I know them.

However, the standard interpretation of quantum mechanics states that *what can't be measured doesn't exist*. In this view, the contents of my fridge would not even exist until I opened the fridge door. In the early days of quantum theory, Niels Bohr was the most ardent proponent of this viewpoint. When quizzed by a fellow physicist on what state a particle must *be* in, Bohr replied in exasperation 'Be! Be! But what is this be?' Bohr claimed that 'an independent reality in the ordinary physical sense can neither be ascribed to the phenomena nor to the agencies of observation.' He was so persuasive that this viewpoint became enshrined as the standard interpretation of quantum mechanics, prevailing for much of this century. It is often known as the Copenhagen interpretation in recognition of the influence of Bohr and his co-workers at Copenhagen's Institute of Physics, where many of the fundamental advances in quantum theory were made.

In the Copenhagen interpretation of quantum mechanics, properties such as position and momentum become real only when measured. This is a revolutionary concept, but in itself amounts to little more than late night metaphysical ramblings concerning whether a tree makes a sound or not when it falls in the forest. I leave such thorny problems to future generations of undergraduates. What is of much more relevance is that, in quantum mechanics, measurement not only makes reality, but that measurement inevitably *modifies* reality.

You will recall that the uncertainty principle defines a minimum value for the product of the uncertainty of position and the uncertainty of momentum – but neither one alone. In any experiment we can choose to measure one of these complementary properties to a high degree of accuracy. If we choose to measure position, the observations we make of a particle's position will make position a real property whilst momentum remains uncertain and *unreal*. Similarly if we measure momentum, our observations would make the particle's momentum real, but at the cost of sacrificing reality for its position. In quantum mechanics, measurement is never innocuous: it modifies *the reality of what is out there*. Measurements make reality and different measurements make different realities.

It may, however, seem to you that abandonment of the conventional notion of external reality seems a steep price to pay for our observations of the pattern of light emerging from a hole. Indeed it is. We have so far demonstrated only that when we attempt to measure the position of a photon of light its momentum becomes uncertain; and vice versa. But though the world may be constructed in such a way that it precludes us from ever measuring both position and momentum simultaneously, does that really force us to abandon the notion of the reality of a particle's position? Why should we slavishly adopt the Copenhagen interpretation? Surely there are more conventional explanations of these experimental results? To examine why any interpretations of quantum mechanics must incorporate a radical view of reality, we must boldly go further than our simple experiment of shining a light through one hole: and shine a light through two.

LIGHT AND TWO HOLES

Consider a beam of light but one shone through a screen with a one-centimetre-wide slit. If we choose to measure momentum for a photon passing through the slit, then its position becomes uncertain and, according to the standard interpretation of quantum mechanics, *unreal*. But let us for the moment deny this viewpoint and (with clenched fists) assert our photon's right to occupy a real position within the slit, irrespective of whether or not we choose to measure it. If a single photon does occupy a real, unique position, then it must pass through one discrete area of the slit but not others. To prove the reality of position for our photon – our *realist position assertion* – we will replace the screen with a single wide slit with another pierced by two adjacent narrow slits, therefore forcing our photon to choose a particular path through the screen, either through the left or the right slit (FIGURE 7.2).

We first perform the experiment with the left slit covered up. This is essentially the same set-up as the single-slit experiment and we see a diffuse (diffracted) image upon the wall (labelled [RIGHT] in FIGURE 7.2). Similarly, with the right slit covered up we will see a similar diffuse strip of light (labelled [LEFT]). With both holes, we might expect to see the product of the two single diffraction patterns, [RIGHT] + [LEFT] (FIGURE 7.2). This is the pattern we would expect if firing conventional particles (such as bullets) through the slits. But when we shine light through the screen we do not see the expected composite pattern but instead a series of light and dark bands on the wall (FIGURE 7.2). This image is certainly not a simple addition of two independent beams of

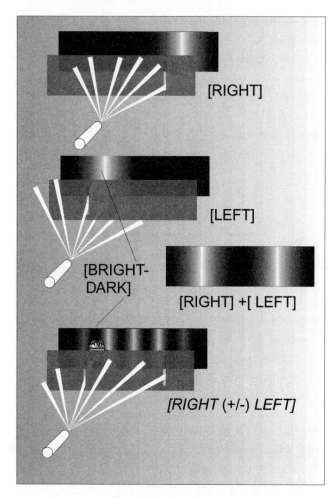

FIG 7.2   Two-slit experiment. When a beam of
light is shone through two slits, an interference
pattern of light and dark bands is generated on the
screen. This interference pattern is not a simple
sum of two diffraction patterns. [RIGHT] and
[LEFT] show the diffraction patterns for light
passing through either the left or right slit;
[RIGHT] + [LEFT] is the sum of these patterns;
whereas [RIGHT (+/−) LEFT] represents the
image (interference pattern) obtained when light is
passed through both slits. The area
[BRIGHT-DARK] is bright when only the left
slit is open, but dark when both slits are open.

light, since areas on the wall that were bright with either single slit open are now dark and other areas that were dark are now bright. This pattern of light and dark bands is known as an *interference pattern*.

Imagine yourself standing against the wall at the position marked BRIGHT-DARK in the Figure (ignore the frog for the time being – I will be introducing him later). From this position, in a dark strip of the interference pattern, no light reaches you from either slit, so it looks as if neither is open. Yet you can confirm that they are indeed open by taking a step to the right or to the left (into a bright band), when from your new viewpoint you can see that both slits are bright. Now (standing within the dark band), you ask a colleague to close the left slit. Immediately you see a bright strip of light shining out of the right slit. Yet all that has happened is that the left slit has been covered! How can preventing light passing through the left hole, cause light to emerge from the right?

The reason is that, at your position, light can reach you from the right slit, but only if the left slit is closed. With the left slit open, the light emerging from that slit interferes with light coming from the right. So you can see that the interference pattern is an odd kind of addition of the patterns of light you see when either slit is open. At some positions, the light from both sources adds up to generate a bright band; whereas, at other positions, the light from each cancels the other out to generate a dark band. To indicate the anomalous nature of the pattern of light and dark strips we will designate it as [*RIGHT (+/−) LEFT*]. The (+/−) term and italics denote that the interference pattern is a complicated addition of the image with both slits open (+) and that obtained with neither (−).

This is, in fact, more or less the same as a classic experiment of Thomas Young in the early nineteenth century. It was one of the crucial experiments that 'proved' the wave theory of light. Light, so the theory goes, is emitted as a series of spherical wavefronts. When the waves pass through the screen of the two-slit experiment, the emerging light from a single slit becomes the source of its own series of spherical wavefronts. The pair of wavefronts emerging from the pair of slits converge and *interfere* with one another before they reach the screen. The light bands correspond to regions where the two waves arrive *in phase* (peaks and troughs marching together) and reinforce each other. The dark bands are areas where the waves arrive *out of phase* (peaks meet troughs) and cancel each other out.

Interference is a feature of waves. Waves are spread out across space, allowing them to pass through two places at once – the two slits in our screen – but then to be recombined to generate the interference pattern. Interference is not a feature of classical particles. Classical particles are localized in space and cannot travel through two separated points. It was Young's discovery of light interference effects that clinched the wave theory of light for nineteenth-century physicists.

But how then can we square interference effects with the evidence for particles of light, photons? How do we account for phenomena like black body radiation, the photoelectric effect and even the silver dots making up a photographic image? The simplest interpretation would be to conclude that the wave-like behaviour of light is a manifestation of the dynamics of millions of photons. In this case, single photons would have *real* positions in space and their individual dynamics would betray no trace of the wavy interference patterns. To prove our *realist position assertion* for photons, all we need to do is to repeat the two-slit experiment, a photon at a time.

To test this idea, we must again reduce the light intensity so that only single photons are emitted, one at a time, from our source and are individually detected at the photographic plate. If we do this with only the right slit open, the collection of dots fall within the same area [RIGHT] illuminated by the strong beam of light diffracted as it passed through the right slit. We will call this distribution of photon landings a *scatter* pattern, as it is exactly the distribution we would expect for any particles (such as bullets) scattered as they passed through a narrow opening. Similarly, with the left slit open, the photons land within the illuminated area [LEFT] that represents the scatter pattern for particles going through the left slit. With both slits open, we might expect that individual photons must now pass through either the left or the right slit. For any photon that *chooses* to pass through the right slit, it should make no difference whether the left slit is open or closed. We would expect it to fall within the same [RIGHT] area as it did when the left slit was closed and this was the only route open. Similarly, any photons that *chose* to travel through the left slit, we would expect to land within ([LEFT]). The pattern of dots would then form the ([RIGHT] + [LEFT]) image that is a simple product of the scatter patterns for each single-slit experiment.

We fire one photon at the screen with both slits open and, as expected, the photon's arrival is recorded as a discrete silver dot on the photographic plate. We fire another and get a second dot. We keep on firing individual photons until there are lots of dots on the screen. Did each photon pass through only one slit in the screen? As the pattern of dots recording each photon landing accumulates on the photographic plate, the image obtained forms exactly the same interference pattern of light and dark bands [RIGHT (+/−) LEFT] that we saw with the strong beam of light. No scatter patterns: single particles show interference effects.

Our aim was to demonstrate that position was a real property of photons (although it may not be simultaneously measurable with momentum). If we were right, then light particles, taking a single defined trajectory through space, should have travelled by a single route through only one of two slits. Yet, single photons appeared to pass through both to generate an interference pattern.

Photons must, like a wave, pass through both slits at once. The conclusion is inescapable. Contrary to our *realist position assertion*, individual photons do not always have a real position in space. Quantum mechanics wins.

Richard Feynman, Nobel prize-winning physicist and architect of the enormously successful theory of quantum electrodynamics (QED), considered that 'the experiment with two holes . . . has in it the heart of quantum mechanics. In reality, it contains the only mystery . . . In telling you how it works we will have told you about the basic peculiarities of quantum mechanics.'[3] The experiment with two holes reveals the wave-particle duality of both matter and radiation. Photons may be emitted as particles, they may be detected as particles, but when unwatched, they travel through space as waves. This wave-particle dualism of matter is described in quantum mechanics as *quantum superposition*. When a photon travels through the screen, the photon is said to exist as a quantum superposition of a photon going through the left slit *and* one going through the right. Neither possibility (photon passing through right or left slit) is entirely real on its own. Quantum superposition is a basic ingredient of quantum mechanics. It may, as we have seen, refer to a superposition of position states (a particle in two places at once) but it can equally describe a superposition of momentum states, energy states, angle of polarization, angle of spin or indeed any property of a quantum system.

Wave-particle duality, quantum superposition and the uncertainty principle are, in fact, different ways of expressing the same underlying quantum-mechanical reality. When we measure momentum for a particle, the uncertainty principle forces its position to become spread out in space like a wave. Its position exists as a quantum superposition of all possible position states. Similarly, when we measure position for a particle, the uncertainty principle forces its momentum to take on the character of a wave and exist as a quantum superposition of alternative momentum states.

All matter exhibits this curious mixture of wave and particle character and the uncertainty principle describes the relationship between these complementary properties. At the back of the tube of your television is an electron gun that fires a beam of electrons at the screen. If the beam is squeezed through a pair of narrow holes, an interference pattern is seen on the screen. The interference patterns demonstrates that electrons, like photons, travel through space as waves and may go through two holes at once. Yet discrete pinpoints of scintillation – the hallmark of particles – herald the arrival of single electrons at your TV screen! Experiments similar to the classic two-slit experiment have been performed with all manner of particles: protons, neutrons, atoms, molecules and ions: they all show interference effects; all demonstrate the same wave-particle duality; they may all exist as a *quantum superposition of states*. The most remarkable demonstration so far has come from the Experimental Physics Group at Vienna University who observed interference effects for C_{60} fullerene, the famous 'buckyball, a big molecule with 60 carbon atoms. Somehow each complex molecule passes through both slits in the two-slit experiment as a single coherent undivided wave/particle.[4] And these particles – electron, protons, atoms and molecules – are not part of some esoteric state of matter with only a fleeting existence in a superconducting supercollider or other high-energy physics experiment. These particles are the stuff of the page you are reading, the eyes through which you are seeing it and the brain with which you are conscious of what you are seeing. Quantum mechanics describes the fundamental reality of matter, including the matter of life.

But it still seems very odd. How is it that atoms and molecules, with their individual constituents of protons, neutrons and electrons – particles that we can see (with powerful electron microscopes) to have discrete sizes and shapes – how can these complex entities somehow *unwrap* themselves to travel through space as a diffuse wave passing through distant points simultaneously? The uncertainty principle, wave-particle duality and quantum superposition are fundamental to quantum mechanics, yet remain one of science's strangest aspects.

Quantum mechanics, as described so far, is very much in the terms of the official version, developed largely by Niels Bohr and colleagues – the Copenhagen interpretation. I will describe a number of alternative interpretations in Chapter Nine. Each is bizarre in its own way; none restores classical reality as it was before Planck. Some of the alternative interpretations do, however, retain standard conventional particles at all times, but must then account for interference effects by other unconventional phenomena such as backward-in-time-signalling between particles. Each interpretation retains the basic features of quantum mechanics and each predicts the same results for experiments. The peculiarities of quantum mechanics are still there, but are wrapped up in different kinds of packaging.

The uncertainty principle may appear baffling, yet its consequences are profound. The early models of the atom stated that the negatively charged electrons *should* crash down upon the positively charged nucleus. What keeps electrons up? The uncertainty principle. If we imagine an electron that has fallen into the nucleus, then its position would be confined to the tiny radius of the nucleus. This confinement in position invokes the uncertainty principle and must be balanced by a huge degree of uncertainty for the particle's momentum. That extra momentum *borrowed* from the uncertainty principle would immediately cause the electron to fly right out. The electromagnetic force between electrons and protons is insufficient to hold the electrons within the volume of a nucleus (in case you are wondering how protons and neutrons are held together within the nucleus, it is due to the far stronger *strong nuclear force* between them). The quantized electron shells or orbitals are a compromise between the demands of the electromagnetic forces in keeping the electrons and protons as close as possible and the constraints of the uncertainty principle prohibiting them from getting too close.

The lowest energy state for a single electron orbiting the single proton nucleus of a hydrogen atom can be roughly estimated simply by calculating the minimum energy allowed by the uncertainty principle for a particle confined in its position to a box the size of an atom. This lowest energy state defines what is known as the ground state (or lowest electron shell). The higher energy states are multiples of this minimum energy. The quantitization of energy states for electrons within atoms – the basis of all chemistry – is therefore founded on the uncertainty principle. Atoms *need* the uncertainty principle to exist. This fact more than any other demonstrates that the uncertainty principle represents a fundamental property of matter, rather than merely a limitation of our powers to perform delicate measurements. Atoms existed long before we constructed measuring devices. Electrons stayed out of the nuclei billions of years before we were here to record their feat. The uncertainty principle lies behind a multitude of other physical phenomena, from the burning of fuel inside the sun to the hydrogen bomb, to the action of lasers and the hardness of metals.

We now have in our possession most of the basic ingredients of quantum mechanics: the uncertainty principle, complementarity, wave-particle duality and superposition of states. Armed with these, we must reconsider the question that made us turn to quantum mechanics: what makes things move? In Chapter Five, we discovered that enzymes do their job by directing particles along particular paths in the cell. For instance, the reaction catalysed by the enzyme LDH is initiated by the directed motion of a proton along a path between a histidine amino acid (on the protein) and the lactate substrate molecule. At any moment in time we might imagine the proton as either having made its way over to the substrate or remaining attached to histidine. But a proton is a fundamental particle subject to Heisenberg's uncertainty principle, and as such, must always be described as a quantum entity capable of existing in a superposition of position states. In reality, the proton must occupy both locations – stuck to the histidine amino acid *and* attached to the lactate molecule – as a quantum superposition. Similarly, when we considered before a mutation that may be initiated by the motion of a proton from one site in the DNA molecule to another (tautomerism), quantum mechanics tells us that the proton exists in both position states at once as a quantum superposition.

But then how does anything ever happen? Whenever we move our arm or leg, protons must jump from myosin-bound water molecules to ATP (as described in Chapter Five) to initiate muscle contraction. But quantum mechanics insists that the complementary non-jumping events must also exist as part of the quantum superposition. What happens to those complementary events? Imagine a professional footballer, poised to take a penalty kick. His leg muscles might be in such a critically balanced state that a single quantum event – a proton jumping (or not) from myosin to ATP – could modify the ball's trajectory just enough to make it veer a millimetre from the goal-mouth. But if the proton exists as a quantum superposition of having jumped or not, then the striker (and his legs) exist as a quantum superposition of having scored the goal or not. The football must exist as a quantum superposition of a football flying into the net and of another hitting the crossbar and bouncing back out again. The crowd of supporters is condemned to a superposition of cheering for joy and gasping with disappointment. If the game was the last of the season for a relegation-threatened side then the presence/absence of the goal will be associated with a superposition of very different life histories for the entire football team, their management and fans.

Quantum events which cause mutations must have had an even more pervasive influence on history than football goals. The origin-of-life scenarios in Chapter Four must have involved a multitude of quantum events. The subsequent appearance and development of life from the earliest cells through to vertebrates and ourselves must have involved trillions of mutations, each contingent upon a quantum event that might or might not have happened. Quantum mechanics tells us that at some level, each mutation both did and did not happen as a quantum superposition. In his book *Wonderful Life*, Stephen Jay Gould argues that the evolutionary history of life on Earth has been contingent upon a multitude of chance events. Play the tape of life again and an entirely different evolutionary history would unfold, with every likelihood that humanity would not be part of it. Many of these evolutionary contingencies must have been quantum level events. If the particles which cause mutations exist as a superposition of position and energy states, then the animals that inherit those mutations would exist as a superposition of mutant states. The entire history of life on Earth would become a massively complex superposition of every possible animal, plant and microbe that could ever have existed.

Clearly footballers and other animals do not live in a superposition of states. But equally clearly, their protons, electrons and photons do. Something has to give.

8

Measurement and Reality

Despite a multitude of experimental 'proofs' of quantum mechanics, there appears to be an obvious and fundamental flaw with a theory involving quantum jumps, quantum superpositions and complementarity. Quite simply, the world we see, hear, feel, taste or smell does not adhere to the rules of quantum mechanics. We never *see* a single goal that has been both scored *and* not scored; we never see animals that have mutated *and* not mutated. We never see quantum jumps and have no problem in measuring both the position and momentum of a car or a billiard ball. Why not? The first and most obvious source of this dichotomy is the size of the objects concerned. Quantum mechanics applies to the microscopic constituents of matter: electrons, photons and atoms. The *real* world we see is big, composed of enormous numbers of these particles. The actions we are used to seeing, such as the flight of a kicked football or the flow of a river, are generated by the movement of billions of particles.

Classical physics was, and remains, very successful in describing and predicting the behaviour of the teeming mass of particles in bulk matter. It was only when its predictions clashed with the findings of experimental physicists that it became necessary to construct new tools for dealing with the microscopic world, and so quantum mechanics was forged. The first rivet hammered into the theory was Planck's quantum of action (involving $\hbar$) and we have already encountered one distinction between the world of the big and the small, in the relative importance of $\hbar$. Quantitization on the tiny scale of Planck's Constant is irrelevant for large bulky objects, but becomes vital in accounting for the dynamics of fundamental particles, including those inside living cells. Is then the size of Planck's Constant sufficient to account for the absence of quantum rules in our everyday reality? No, for there is much more to quantum mechanics than the smallness of the quantum of action. The principle of quantum superposition of states knows no size limit. It is perfectly valid in standard quantum mechanics to describe my being in two places at the same time or doing two things at once.

Why can't quantum superpositions of states be experienced in the *real* world? Single photons are able to travel by two different routes before arriving as a single particle at a screen. Then why can't we drive both ways round the block to reach the cinema? Have you ever taken a shortcut to avoid busy traffic only to find stuck in another jam, and been left wondering whether you would have been better off sticking to your original route? You are always left wondering. But quantum mechanics could find an answer to this conundrum. You would simply drive along both routes simultaneously as a quantum superposition and discover which version of *you* arrived first in the queue at the hot dog counter.

The explanation favoured by Niels Bohr emphasized the distinction between the quantum world and *measuring devices*. To see quantum events we must amplify them to the level of big classical objects, if only because, 'the account of all evidence must be expressed in classical terms'.[1] Bohr pointed out that our senses of perception and powers of communication are rooted in classical laws. It is impossible for us to either perceive or describe quantum events except in terms of classical phenomena. All our interactions with the quantum world must be translated into classical *language* by some kind of interpreter. Measuring devices are the interpreters which convert quantum events to classical phenomena: the position of a pointer on a dial, the click of a Geiger counter or indeed any detail of the world around us.

The world is thus divided between an invisible quantum domain and the familiar realm that we perceive, obeying classical laws. Straddling that border are quantum-measuring devices. Physicists who examine the quantum realm utilize various measuring devices, ranging from the simple photographic plate to the twenty-seven kilometre long CERN Large Hadron Collider beneath Mont Blanc. Yet the nature of these tools, the measuring devices, remains the most enigmatic aspect of quantum mechanics. I will be arguing that, at its most fundamental level, the living cell is a quantum-measuring device which measures its own internal state.

To understand life we must therefore examine how quantum-measuring devices make the world *real*.

QUANTUM MEASUREMENT AND INFORMATION

Let us return once more to the famous two-slit experiment (as you can now see, Feynman was not exaggerating when he said that the experiment with two holes contains 'all the basic peculiarities of quantum mechanics'). Our original aim was to accurately measure position and momentum for a photon – or electron or any other particle. We found that our ambition to achieve simultaneous measurement of both quantities was limited by the uncertainty principle. Our (less ambitious) revised aim was to demonstrate that the attributes of position and momentum were in fact real (though not simultaneously measurable) for a photon. This amounted to attempting to demonstrate that a single particle travels by a single route through only one of the slits to form scatter patterns. Yet single photons appeared to pass through both, generating an interference pattern.

But, perhaps, we remain unconvinced and wish to catch our photons during their travels and directly see them performing their trick of travelling by two routes. At present we have only a single measuring device within the experimental set-up, the photographic plate. However, its measurement does not amount to a position measurement for photons at the slit screen, since photons travelling by either route might reach the same point on the plate. To catch our photons *en route* we will have to add an additional quantum-measuring device. A suitable one would be a photon-detector able to detect the passage of a photon during its flight[2] and amplify the signal to give an audible beep. As usual, we will start with a strong visible beam of light, initially placing our photon-detector just downstream of the light source (the torch or laser pointer). At this position the photon-detector cannot distinguish between the possible routes open to the photon but can investigate whether the light it *sees* is a particle or a wave. Our photon-detector initially registers the emission of a bright beam of light as a continuous beep. But if we reduce the light intensity to a level at which black dots are formed one at a time on the photographic plate, the photon-detector emits single discrete beeps for each photon-detecting event. The additional quantum-measuring device has thus succeeded in confirming that our source does indeed release light as a stream of particles. The photographic plate records the arrival of particles. Yet despite both these particle measurements, the black dots that accumulate on the photographic plate form the same [RIGHT (+/−) LEFT] pattern of light and dark bands as before: interference. Once again, light is emitted as particles and arrives at the photographic plate as particles, but betrays the evidence of an intermediate wave-like existence by forming the interference pattern.

We are, however, determined to discover precisely where in the experimental set-up the transition between particle and wave behaviour occurs; so we purchase an additional pair of photon-detectors and place one at each slit. If, as the interference pattern suggests, individual photons pass simultaneously through both slits then we would expect to hear both detectors beep in unison (with perhaps half a beep each) for each passage of a photon through the screen. When we repeat the experiment with both detectors what we actually observe is that for each photon that passes through the screen, only one detector beeps. The left detector beeps for about half the photons and the right for the remainder. We never hear both detectors registering the passage of a single photon. Individual photons appear to travel through only one slit at a time, never both. Surely it is particles after all, not waves, that are passing through the slit?

But wait. When we now examine the photographic plate, the interference pattern has vanished. Instead of the complex [RIGHT (+/−) LEFT] series of light and dark bands, the accumulated dots now form (Figure 7.2) a simple sum of the two diffraction patterns ([RIGHT] + [LEFT]). But the wave theory of light was built upon the observation of those light and dark bands. With no interference pattern, there is no longer any need to propose a quantum nature for light. Photons now behave as sensible particles at the source, at the slits and at the photographic plate. What has happened to their waviness?

Our first thought might be that the extra measuring devices have disturbed the light waves passing through the slits and destroyed the interference effects. We can test this hypothesis by removing the extra two detectors and positioning the original detector so that it now sits above the right slit. From the earlier experiment, we know that the operation of this single detector failed to disturb the interference pattern when placed downstream of the source. But from that position, it was unable to distinguish between photons taking the right or the left path. In its new vantage point, the detector will be able to distinguish photon paths but should perhaps leave the interference pattern intact. Yet, when we repeat the experiment, the dots once again accumulate within the scatter pattern ([RIGHT] + [LEFT]) we would expect for the sum of two diffraction patterns. No interference effects! It seems that the single detector is sufficient to destroy the wave character of light, but only when it can distinguish between photon paths.

The waviness of light seems to be sensitive – not so much to measurement *per se* – but only to those measurements that extract 'which way?' information. When the detector is placed where it cannot *see* the photon path, the quantum superposition of position states – the photon's waviness – persists. But if that same detector is placed where it can distinguish paths, then the waviness collapses to yield particles: photons. It is at this point that the experimental system is said to perform *a quantum measurement of position* on the photon. The quantum measurement destroys quantum superposition and abolishes the interference pattern. In standard quantum mechanics, we say that the quantum superposition [*RIGHT (+/−) LEFT*] *collapses* to yield either of the simple states: photon going through right slit [RIGHT] *or* photon going through left slit [LEFT]. Upon measurement, the individual photon must now *choose* whether it will travel through the left or the right slit: [*RIGHT (+/−) LEFT*] + *Qmeasurement* → [RIGHT] or [LEFT]. This *information extraction* aspect of quantum measurement will be vitally important when we come to examine the action of living cells as quantum-measuring devices.

FROGS SEE

Another counter-intuitive aspect of quantum measurement, that it is non-local, becomes apparent if we place an observer at the level of the photographic plate. Since our eyes cannot see single photons arriving, we might recruit the help of a reliable and conscious (this *may* be important

– see the next chapter) frog physicist (remembering that frog's eyes are sensitive to individual photons, although whether they can actually *see* them is of course much more difficult to answer) to record the sightings of photons. We ask our curious frog physicist to sit at the position labelled *BRIGHT-* in front of the wall. As before, *BRIGHT-DARK* was within one of the dark interference bands when both slits were open, but is an area that was bright with either slit open. We first open only the right slit and our frog reports seeing individual flashes of light (photons) emerging from the right. We then cover the right slit and open the left slit and our frog duly reports flashes of light emerging from the left. Finally, we open both slits and our frog at A is plunged into darkness (remember he is located within a dark strip of the interference pattern) and reports no sightings of light flashes from either slit.

Our frog standing with his back to the screen is not aware of any interference pattern. All he sees is flashes of light emerging from one or other of the slits. He is therefore very puzzled that whilst a moment ago he saw flashes of light emerging from the left slit, now that we have opened both, he sees no light. 'What's going on?' he exclaims. 'How can opening up that hole on the right prevent light from getting through the hole on the left?' You reassure him that this is all perfectly compatible with the quantum-mechanical nature of light but warn him that there is worse to come.

With both slits open (and interference pattern intact) we next place a photon-detector over the right one. The photon-detector performs the quantum measurement: [*RIGHT* (+/−) *LEFT*] + *Qmeasurement* → [RIGHT] *or* [LEFT]. This measurement abolishes the lights waviness and forces the photons to choose either right or left slit. The interference pattern disappears and we are back to the simple [RIGHT] + [LEFT] diffraction pattern. Remember that point *BRIGHT-DARK* was bright when either slit was open, so, in the absence of interference effects, point *BRIGHT-DARK* will also be bright with both slits open. Our frog accordingly reports flashes of light but now coming from both slits. He also reports puzzlement that a measuring device placed at R could cause light to emerge, not only from R, where the detector is positioned, but from L, a place distant from the detector.

Our sceptical frog suspects trickery, so asks us to switch the detector at R on and off. When the detector is switched off (allowing the interference pattern to form) he sees no light from either slit; but with it on he sees flashes of light from both slits. What our frog finds most perplexing is that those photons he sees emerging from the left slit cannot have been detected by the detector placed at the right. Whenever a photon passes through the left slit, it clearly has not passed through the right slit. *The detector at the right slit cannot be detecting any of these events.* Yet the frog sees photons of light emerging only from the left slit when the right detector is switched on, *but recording nothing*! When the detector is switched off, *and still recording nothing*, no light emerges from the left slit. The mere presence of an (active) detector at the right slit, *though recording nothing*, causes flashes of light to appear from the left slit!

Our, by now, bewildered frog suspects some secret communication. Perhaps the right slit is sending a signal to inform the left whether or not its detector is switched on. The left slit may then be able to adjust its light transmission properties accordingly. 'Aha,' he says, 'I know a thing or two about signals.' In fact, our frog is well trained in classical physics; indeed he is a prince among frog physicists and particularly well versed in relativity theory. He has therefore devised a particularly clever strategy to outwit this putative communication channel. A fundamental principle of general relativity is that signals cannot travel faster than the speed of light. In point of fact, the theory does not prohibit faster-than-light signals but does demonstrate that such signals would have to be travelling backward-in-time. Backward-in-time signalling would wreak havoc with causality: as the future could influence the past. This could lead to all sorts of uncomfortable paradoxes. You could convince your grandmother what a bad lot your grandfather was, and thereby discourage their marriage and so do away with their descendants, including yourself. The prevailing view amongst physicists is that the possibility of generating these paradoxes rules out faster-than-light signalling.

The clever frog instructs us to construct an arrangement of screen, slits and detector, so that the distance between the slits is so great and the switching on and off of the detector is so fast, that no signal – even a signal travelling at the speed of light – could inform the left slit whether or not the right detector is switched on, during the flight of a single photon. Our frog repeats the experiments under these stringent conditions but, alas, witnesses the same result. He only sees light emerge from the left slit when the (by now distant) detector placed at the right slit is switched on. The presence of an active detector at the right slit appears to causes an instantaneous (or at least, faster than light) change in the properties of the left slit, though it detects nothing.

Determined not to be outwitted, the frog devises another ingenious modification to the experimental design. He now instructs us to place the photon-detector *downstream* of the right slit (between the slit and the photographic plate), though sufficiently within the *slipstream* of the slit that it can still unambiguously detect a photon that has passed through the right slit. Surely now, with the detector positioned so that it detects photons only after they have passed through either slit, it can no longer influence their passage through the left slit?

But it still does. The results are precisely the same as when the detector was placed directly before the slit. The frog only sees flashes of light coming through the left slit, when the detector is switched on. But now the detector can detect only photons that have already passed through the screen. The state of the measuring device (whether switched on or off) seems to be capable of influencing an event that has already happened.

The puzzlement experienced by our frog (and perhaps by the reader) is echoed by Einstein's incredulity; he described the *non-local* aspect of quantum measurement as 'spooky action at a distance'. Yet this spooky action is undoubtedly an element of reality. The next chapter will describe real experiments that have confirmed the reality of these phenomena.

DOING NOTHING

One feature of quantum measurement perhaps even more bizarre than non-locality is the ability of measurements that record nothing to influence the system's dynamics. In the two-slit experiment, the presence of the detector at the right slit, though it measured nothing, still influenced what our frog saw at the left slit. Physicists describe this phenomenon as a *null measurement*. Though the right slit detector measured nothing, that null measurement was nonetheless a quantum measurement that can propagate its influence at superluminal speed (faster than the speed of light) to the left slit.

To help us understand these peculiarities, let us utilize the concept of the quantum wave function, designated by the Greek letter, ψ. The wave function is in reality a very complex mathematical object but we will use it solely as a kind of shorthand to help describe the quantum world. In the two-slit experiment, when a photon is released by the source, we will describe it with a quantum wave function, ψ [photon]. When the photon passes undetected through the screen then its wave function passes through both slits as a superposition of both position states: ψ [{photon at RIGHT slit} (+/−) {photon at LEFT slit}]. The wave function has many properties of a classical wave and can, like a wave, pass through two or more points simultaneously. However, a photon-detector placed at the right slit will perform a quantum measurement to *collapse* the wave function and destroy the superposition of position states: ψ [{photon at RIGHT slit} (+/−) {photon at LEFT slit}] + *Qmeasurement* → ψ [{photon at RIGHT slit detected by R detector] *or* ψ[photon at LEFT slit not detected by R detector]. The alternative states measured by this device (which beeps to record the passage of a photon or does not beep to record the absence of a photon) are termed *eigenstates* of the measuring device (eigenstate means 'own state' in German). Measurement forces the quantum system to *choose* between the possible eigenstates of the measuring device. After measurement, the quantum system plus the measuring device must be found in one of these eigenstates.

Now it is obvious that it does not matter whether the detector is placed at the right or left slit, because it is the wave function ψ that encounters the measuring device and travels through both. After measurement, the wave function collapses and the resulting photon may be found at either slit. The measurement performed at the right slit collapses the wave function for the photon everywhere (at every possible position) instantaneously. This is the source of the mysterious communication between right and left slit − the collapse of the wave function over all the space it occupies on encountering the measuring device (spooky action at a distance). Null measurements are only paradoxical if we persist in thinking of photons existing as real particles inbetween measurements. If we accept instead the reality of a non-local wave function for the inbetween measurement states, then the paradox disappears. The null measurement is merely one of the two possible eigenstates of the measuring device.

A WORLD WITHOUT CAUSE

We said earlier that, after a quantum measurement interrogates the wave function, the photon may be detected in either left or right slit. If the two-slit experiment has been set up symmetrically, then half the photons will be found in the left slit and the remainder will have travelled through the right. But what if we have only a single photon? It will have a fifty per cent chance of being found at the left slit and a fifty per cent chance of being found at the right. Say it is detected at the left slit. We may ask – what caused it to travel through the left slit rather than the right? If this were a classical system, then the photon's precise trajectory would be *caused* by tiny variations in how the photon was fired, and perhaps also by the minute interactions with the atoms or molecules it encounters on its travels. All these tiny variations would add up to a kind of *pseudorandomness* in which the photon would sometimes be knocked one way, sometimes another, and would end up thereafter in the right or left slit with equal probability. However, there would always be some way of reducing this randomness – perhaps performing the experiment in a vacuum or firing the photons from a narrower source. Eventually, with enough careful control of the conditions it would be possible to eliminate this pseudorandomness. It should then be possible to fire a single photon, determine through which slit it passed, repeat the experiment, and obtain the same result. Randomness would be eliminated. The photon's precise trajectory would have a *cause*.

It is, however, a fundamental tenet of quantum mechanics that events at the level of fundamental particles do not have causes. What decides which slit the photon will be found in after measurement-induced collapse of the quantum wave function? Nothing! It is entirely random. In quantum mechanics it is *in principle* impossible to set up the two-slit experiment in such a way as to be able to predict through which slit a single photon will travel, however carefully you control the conditions. With even the most stringent control, the best you will ever achieve is an equal probability for the photon being found in either slit. This leads us to the remarkable conclusion that in this system (and in any other quantum system) there are no causes to separate the possible alternative events. Quantum events simply happen. This represents a fundamental break with that bedrock of classical science, causality – the principle of cause and effect. Here we have an effect – photon travelling through the left slit – which has no cause. This lack of causality is what makes quantum mechanics *non-classical*. This troubled Einstein so profoundly that he could never entirely accept the truth of the theory he helped to establish.

REALITY AS PROBABILITY

Although in quantum mechanics, it is impossible to predict precisely where an individual photon will land, it is possible to make predictions of the likely distribution of large numbers of photons: you can calculate the *probability* of a photon landing in a particular area. Calculations of this type of probability distribution are what most of quantum mechanics is about. The key parameter used is the wave function, ψ. In principle (though not in practice) the calculations are simple. Max Born, a founding father of quantum mechanics, demonstrated that classical probabilities are obtained by squaring the *amplitude* of quantum wave functions.[3] Amplitude is a wave property: it describes a wave's height. The amplitude of the wave function can be described as the height *of what is waving* in the wave function. So, to calculate the probability of a photon landing at any point on a photographic plate, we merely need to know the amplitude of the wave function at that point and multiply it by itself to obtain the probability for a photon landing.

With just a single slit open, say the right, there is only one possible route for photons; it is simple to calculate the amplitude of ψ at any point on the photographic plate and to square that value to obtain the probability for a photon landing. If we plot the probability density across the plate as a series of white dots for high probability and black for low, then the dots will reproduce the simple diffraction pattern.

When both slits are open, the calculations are more complex. Now we must take account of the two different routes that the wave function can take before it arrives at the plate. Just like a real wave, ψ splits when it meets the slits, into two wave function *beams*: ψ_L and ψ_R. Once the beams reach the plate, their amplitudes must again be squared to convert them into classical probabilities. However, before we can do this, we must first add together the amplitudes of the wave functions that have arrived from the left or right slits: ψ_L and ψ_R. But remember that ψ describes a wave, and waves can be *in phase* or *out of phase*. At some points on the plate the waves will arrive in phase to give high combined amplitude; at other points they will arrive out of phase, cancelling each other out. It is this summing of positive and negative amplitudes that generates the interference effects. The summed amplitudes are once again squared to generate probabilities that we can plot as black (low probability) and white (high probability) dots. The pattern of dots will form a series of light and dark strips on the screen that will reproduce the interference pattern.

Now what happens if we try to detect the passage of a photon? With both slits open, but with a photon-detector placed at one, then the ψ wave function cannot survive intact as far as the photographic plate. Instead, as we have come to expect, it collapses when it meets the detector positioned at the slit. Detection generates a classical signal, so it is at this point that the wave function amplitude must be squared to yield a probability. The photographic emulsion will then encounter a photon (rather than a wave function) that has already *decided* whether it has passed through the left or the right slit. To obtain the probability distribution for photon landings it is still necessary to add together the probabilities for the alternative routes. However, the addition now involves probabilities *not* amplitudes. Probabilities are always either zero or positive, never negative (the *square* of a negative amplitude will still be positive). With no positive and negative numbers to cancel each other out, the interference terms disappear and the pattern of probability dots becomes a simple addition of the diffraction patterns.

It does not make any difference even if the photon-detector is placed downstream. In this case the wave function will remain as a superposition until it encounters the photon-detector. Then the wave function will collapse and amplitudes will be converted into probabilities. But, once again, to calculate the probability for photon landing at any point on the screen, we need to add together probabilities, not amplitudes, and so interference effects are lost.

We can now see that the quantum wave function acts as a kind of square root of the probability for finding a particle at any particular point in space. But the wave function can also be used to describe any other property of a quantum system: energy, momentum, angle of spin or polarization state. The quantum wave function encapsulates all the information it is possible to obtain about a quantum system.

So, to summarize what we have learned so far about quantum measurement. First, quantum systems are described by a wave function that encapsulates all possibilities open to the system, but measurements collapse the wave function and convert quantum-mechanical amplitudes to classical probabilities. Second, to collapse the wave function, it is not enough to interact with a system; a quantum-measuring device must extract some information about the system. Third, quantum measurement causes instantaneous collapse of the wave function everywhere: it is non-local. Finally, a quantum measurement that yields a negative detection – a null measurement – may still collapse the wave function. We have managed to construct nearly all the essential features of quantum mechanics from just two simple experiments.

Quantum mechanics' most enigmatic feature is the nature of the quantum wave function. What exactly is it? It must somehow encompass all possibilities open to a quantum system, and yet when examined all those potentialities vanish, leaving just a single classical reality. How *real* is the wave function in the absence of measurement? What is it made of? What does it *look* like? These are mysteries for which there is no satisfactory answer. Our brains – used to dealing with classical level phenomena – are ill-equipped to form models of the quantum world. The nearest concepts that make any sense are, peculiarly enough, how our thoughts may be undefined until some kind of *measurement* takes place. What do you think of when you hear the word, note? A musical note perhaps, or maybe a written note or a banknote. But, before you have time to decide which meaning is more appropriate, 'note' has something of each meaning. In a sense they are all there as a kind of superposition: musical note + written note + banknote. This kind of conceptual superposition has some semblance to the nebulous nature of quantum objects. And like quantum objects, the *undecided* concept 'note' is fragile and prone to collapse. If you hear 'book' immediately following 'note' then all rival interpretations instantly evaporate and your mind is left with the notebook kind of note. If instead you hear middle C played on a piano, then the superposition collapses leaving merely the concept of a musical note.

Please do not take the analogy too literally. I am not implying (or at least not in this chapter) that our thoughts are in any way quantum-mechanical. But quantum objects and our thoughts do share a nebulous nature, fragile to any kind of examination.

WHERE DOES THIS LEAVE MOTION?

Armed with our understanding of quantum measurement, let us return once again to consider the motion of fundamental particles inside living cells. At the end of the last chapter, we left the LDH enzyme and its substrate in the uncomfortable position of having the catalytic proton both attached to the enzyme and associated with the lactate, as a quantum superposition. Recalling the wave-particle duality of matter, the proton can be considered as a two-humped wave with one hump of the wave hanging over the histidine residue (of the enzyme) and the second above the lactate molecule. During the course of the enzymatic reaction, the amplitude for these waves changes: the wave (initially large above the histidine molecule) gets smaller, whilst the wave initially small over the lactate molecule gets bigger. If the system stays at the quantum level then this is all that would happen. However, if a measurement is made, then the measurement process must *interrogate* the wave function describing the proton. The wave function will collapse to generate a probability (square of the amplitude) for the proton to be associated with either the lactate *or* the histidine.

If, after measurement, the proton is found with the lactate molecule, then it hasn't really *moved* there – at least not in the way we normally understand movement, as going from one place to another via a series of places inbetween. Instead the proton goes from one place to another by a single indivisible event: a quantum jump. In effect, the proton disappears from the histidine-bound atom at precisely the same moment it materializes at the lactate molecule; another example of Einstein's 'spooky action at a distance'. In quantum mechanics all motion, when looked at hard enough, becomes a series of hops from one place to another.

However, the key point is that the enzymatic reaction performed by the enzyme LDH requires some kind of quantum measurement to *move* the proton from one place to another. Without that measurement, the proton must remain perpetually undecided (in a quantum superposition). All enzymatic reactions involve the motion of single electrons and protons and must similarly be understood in the terms of quantum measurement. The chemical changes that lead to mutations also involve the motion of electrons and protons inside the DNA double helix and must be dependent upon quantum measurement. The motion of fundamental particles is the basis of all actions performed by living cells; and this movement is dependent upon some kind of quantum measurement process.

We have discovered how measurements can influence the dynamics of a quantum system (whether or not interference patterns are generated in the two-slit experiment). What about the quantum measurements that take place inside living cells? Can they influence the dynamics of living processes? To answer this we need to take a closer look at quantum measurement.

LIGHT AND SUNGLASSES

The next aspect of quantum measurement is revealed in a simple experiment that anybody can perform with Polaroid sunglasses. The outcome is startling and gives us a new way of looking at quantum measurement, which will be very useful when we examine how quantum mechanics may influence the dynamics of life.

We have already discovered that the choices we make in quantum measurement – which complementary property we wish to measure – influence the future dynamics of a system. This is the essence of the two-slit experiment: our *choice* of whether or not to measure the particle at the slit influences the subsequent *reality*: whether or not an interference pattern is generated. A more subtle illustration of this aspect of quantum measurement can best be seen with polaroid lenses – just like those used in Polaroid sunglasses. If you cannibalize your old sunglasses to yield three polaroid lenses then this experiment is one of the simplest – and oddest – known to science.

We first need to understand what light polarization is all about. Think of a beam of light coming towards you through the page. If the light is polarized, then its waves undulate in a defined direction through the page, say vertically (↕) or horizontally (↔) or at any angle between. For convenience, let us describe the angle of polarization of light according to the time on a clockface such that vertically polarized light is polarized at twelve o'clock (or could equally be described as being polarized at six o'clock) and horizontally polarized light is described as being polarized at three o'clock (or nine o'clock). Most light that we see is not polarized but waving in all possible directions. Polaroid lenses transmit only those light waves that are polarized in a preferred direction – the same direction as the alignment of polaroid crystals inside the lens. They reduce glare because its usual source is light reflected at oblique angles from windows or the surface of water. It is a peculiarity of optics that light reflected at oblique angles is partially polarized and so is filtered out by Polaroid sunglasses.

A convenient source of polarized light is that which has already emerged from a polaroid lens. It is easy to see that light that is polarized at twelve o'clock by its passage through a single polaroid lens is (more or less) one hundred per cent transmitted by a second polaroid lens orientated at the same angle. Conversely, when two consecutive polaroid lenses are orientated so that their axes of polarization are perpendicular (say at twelve o'clock and three o'clock), no light is transmitted through the pair of lenses. Try this with your sunglasses. It is helpful to view a relatively bright object through the lenses, such as a bright light or television screen. Mark the lenses 1, 2 and 3. First find the relative rotation of lenses 1 and 2 that allows the maximum amount of light (brightest image) through. By doing this you are matching the orientation of the polaroid crystals in lens 1 with those in lens 2. Place a mark, such as a vertical arrow, on each to mark this angle of orientation. Repeat this operation for the combination of lens 1 and 3 and similarly mark lens 3. Confirm that all the lenses have the same alignment by allowing light to pass through all three orientated at twelve o'clock when a bright image should again be seen. The order in which the lenses are placed makes no difference to the amount of light transmitted.

Now with lenses 1 and 2, align both their arrows at twelve o'clock to again allow maximum transmission of light. Then, slowly rotate the second lens. As the lens rotates, the image darkens. When the second lens has rotated by ninety degrees, the image completely disappears. This is all perfectly reasonable if we return to our image of light waving either vertically or horizontally. The polaroid lens can be thought of as a kind of fence through which the waves must pass. If the waves are vertical waves then they easily pass through a vertical fence. If the waves are horizontal, they pass through a horizontal fence but are completely blocked by a vertical fence. The combination of a vertical fence and horizontal fence will transmit no light waves, because the horizontal lens will block any vertically polarized waves and the vertical any horizontally polarized light.

Finally, insert the third lens, lens 3, so that it is sandwiched between lenses 1 and 2 but with its arrow rotated to about one o'clock. While no image was visible with two lenses, the image returns with three. Inserting the extra lens somehow allows light to be transmitted through all three lenses. How can the addition of an extra polaroid *lens*, which normally acts only to prohibit the transmission of light, actually promote the passage of light through all three lenses?

Perhaps the polaroid lenses are not acting as we had assumed, to merely transmit polarized light, but are also able to rotate the angle of light polarization. However, another simple experiment can disprove this. Align all three again for maximum transmission of light. If each lens were rotating the polarization angle, their order would matter. The *transmission angle* of one lens would have to match the *capture angle* of the next in line. But, as is easily demonstrated, this is not the case. The order the polaroid lenses are placed in makes no difference to the amount of light transmitted.

The phenomenon is even odder when we consider light as a stream of photons. Can photons be polarized? Indeed they can. I am sure that by now you can guess the experimental set-up necessary to demonstrate this. The polaroid lens experiment is repeated with a low-intensity light source that fires single photons, with a defined angle of polarization, through a polaroid lens. A photographic plate or an array of photon detectors (or frogs) is used to detect the transmitted photons. Photons polarized at twelve o'clock are (more or less) one hundred per cent transmitted by a lens orientated at twelve o'clock, but completely absorbed by a lens at three o'clock. Angle of polarization is therefore a real property for single photons. This in itself is somewhat odd since our concept of polarization is based on an image of a wave undulating in a particular direction. What is waving inside a single photon? How can a single photon carry the directional information of the angle of polarization? What is pointing to that angle? These are questions for which there is no entirely satisfactory answer. As Einstein commented, 'Nowadays every Tom, Dick and Harry thinks he knows it [what a photon is], but he is mistaken.'[4]

With countable photons, we can bring real numbers and quantum-mechanical probabilities into the picture. No photons are transmitted through a pair of crossed lenses orientated at twelve and three o'clock, so we can say that the probability of photon transmission in this situation is zero. However, insertion of a third lens orientated at one o'clock does allow some photons through. In fact, about six per cent of the photons are transmitted through all three. The insertion of an extra lens has increased the probability of photon transmission from zero to six per cent. The phenomenon seems even stranger when considering photons. How can a photon blocked by two lenses be transmitted by those same lenses once a third is inserted between them?

The key to this phenomenon is the realization that the interaction between the photon and a polaroid lens constitutes a quantum measurement. Angle of polarization is a quantum property of photons, like position or momentum. When you examined the light shining through your Polaroid sunglasses, you were in fact performing a quantum measurement. (Anybody wearing sunglasses can be a quantum physicist!) Each polaroid lens is capable of determining the angle of polarization of a photon and performing a quantum measurement. However, unlike the properties of position or momentum, which may have a range of values, a polaroid lens provides a yes/no answer to the question: can light be transmitted through the lens? In quantum-mechanical language, we say that a lens has only two perpendicular eigenstates (the possible states registered by a measuring device). The eigenstate of a polaroid lens with preferred orientation at twelve o'clock are: [photon polarized at twelve o'clock transmitted by lens] and [photon polarized at three o'clock absorbed by lens]. As discovered above, quantum measurement forces a quantum system to adopt one of the eigenstates of the measuring device. Any photon transmitted by a twelve o'clock lens will emerge polarized at twelve o'clock, *irrespective of its previous angle of polarization.* Any photon not transmitted by the lens does not emerge (since it is absorbed) but, if it could be recovered, it would be polarized at three o'clock.

One way of envisaging quantum wave functions which is particularly useful when looking at polarization, is as an arrow in space. In this guise, the wave function is known as the *state vector*, but performs more or less the same role as the wave function used until now. Imagine the state vector of a photon as an arrow that points towards the direction of polarization with length equal to one (arbitrary) unit. Unpolarized light has components that point in all possible directions. A single arrow that points in a defined direction represents polarized light. The state ψ[twelve o'clock] is a vertical arrow, the state ψ[three o'clock] is a horizontal arrow, and the state ψ[one-o-clock] is an arrow rotated 30 degrees from the vertical.

One crucial feature of state vectors is that they can be represented as a superposition of other state vectors which combined, add up to the original vector. For instance, it is perfectly legitimate to write the state ψ[twelve o'clock] as a quantum superposition of two perpendicular states, say ψ[{eleven o'clock} *(+/−)* {two o'clock}], or ψ[{ten o'clock} *(+/−)* {one o'clock}]. Each description of the state vector is exactly equivalent; there is no way of distinguishing the state represented by ψ[twelve o'clock] from a state represented by ψ[{eleven o'clock} *(+/−)* {two o'clock}]. They are the same state.[5] This decomposition of quantum states into a superposition of other states allows us to represent quantum measurement in a new and powerful way. Quantum measurement projects the original state vector onto the eigenstate of the measuring device.

A lens orientated at one o'clock has two perpendicular[6] eigenstates: *one o'clock* (transmitted) or *ten o'clock* (absorbed). When the state vector ψ[twelve o'clock] encounters the lens, it is projected into a superposition of those eigenstates: ψ[twelve o'clock] → ψ[{one o'clock} *(+/−)* {ten o'clock}], with appropriate weighting to take into account the closeness of the original state to each eigenstate. Measurement causes the superposition of the photon plus measuring device to collapse into one of the eigenstates: ψ[{one o'clock} *(+/−)* {ten o'clock}] → either ψ[photon polarized at one o'clock and transmitted by lens] *or* ψ[photon polarized at ten o'clock and absorbed by lens].

Measurement as a *projection* of the state vector onto the eigenstates of the measuring device gives us another insight into quantum measurement. We can consider the state vector as a hidden object we can never *see* directly. All we can do is perform quantum measurements to *project* a shadow of that state onto the measuring device. Different quantum measurements are like different angles of projection: each projects a different shadow onto the measuring device. But there is an additional peculiarity to state vectors; they must be shape-shifters – for once a quantum measurement is made, then the state vector *jumps to become that projection*. The measurement process causes the state of the system to jump to the measured eigenstate. These quantum jumps are another of the non-classical features of quantum mechanics that greatly troubled some early pioneers of the theory. Erwin Schrödinger commented that, 'If one has to go on with these damned quantum jumps then I'm sorry I ever started to work on atomic theory.'[7] Yet despite Schrödinger's antipathy, they remain a fundamental feature of the theory.

Another important point is that quantum measurements are not for ever. If after a quantum system is measured, it is left undisturbed, then it may remain in that measured state indefinitely. However, in reality it will not remain undisturbed for long, as it soon interacts with other particles. Its quantum state will rapidly diffuse into a multitude of possible states as a quantum superposition, until another quantum measurement drags it back to a single state once more. The classical world therefore depends on unrelenting quantum measurement to maintain its reality.

ZENO AND QUANTUM MEASUREMENT

Zeno was born in the Greek colony of Elea in southern Italy in the fifth century BC. He travelled widely for many years, returning to Elea only to be tortured to death after being implicated in a plot to assassinate the city's tyrant Nearchus. He was a resilient character: he is said to have bitten off his tongue, spitting it at the tyrant during torture. However, it is rather for his activities during happier days that he is remembered today; particularly for a series of paradoxes set in the form of short fables. The most famous concerns a race between Achilles and a tortoise. Far slower than Achilles, the tortoise is given a head start of, say, ten metres.

But once the race has started, Achilles easily reaches the tortoise's starting position with a few lengthy bounds. But, by now, the tortoise has moved on. Achilles leaps after it but, on reaching its last position, the tortoise has again advanced by some tiny distance. Achilles again advances, but each time he moves to close the gap it widens. Achilles appears unable to ever reach the tortoise (or even to move at all) because to do so he must advance through an infinity of ever-decreasing distances.

Zeno knew of course that motion was possible and Achilles would catch up with the tortoise. His paradoxes were to illustrate that there must be something wrong with our simplistic notions of time and motion. It took two millennia of mathematical head scratching to solve Zeno's paradox by demonstrating that an infinite sum can add up to a finite number[8]. However, the paradox lives on in the *quantum Zeno effect* and the *inverse quantum Zeno effect*, which describe how a quantum system can be manipulated by measurement.

We have, in fact, already met an aspect of the inverse quantum Zeno effect in our experiment with the three polaroid lenses. You may remember that when two lenses are placed perpendicular to one another (e.g. at twelve o'clock and three o'clock), there is zero probability for photon transmission. However, inserting an extra lens at one o'clock (sandwiched between the first two) increases the probability for transmission from zero to six per cent. What has happened is that the one o'clock lens has projected the state vector onto its eigenstates. Only some of the original twelve-o'clock photons will go through the one o'clock lens, but all that emerge will be polarized at one o'clock. A significant fraction of these one o'clock photons can then pass through the three o'clock lens; while the original twelve-o'clock photons were unable to.

Oblique quantum measurement (measurement slightly shifted from the previous state) is thereby able to rotate the angle of polarization of photons. This process can be continued with more lenses. A fourth lens inserted at two o'clock will cause a further rotation of polarization. The two-o'clock lens will project the state vector of the photons emerging from the one-o'clock lens onto its eigenstates. Only those photons that collapse into the two-o'clock eigenstate will be transmitted, but all those that do will all be polarized at two o'clock. These photons will pass even more readily through the final three-o'clock lens so that the probability for transmission through all four lenses will be increased to twenty-nine per cent. Addition of further lenses will allow even more light through. A series of fifteen Polaroid lenses orientated at one minute, two minutes, three minutes, four minutes and so on up to fifteen minutes, would increase the probability of transmission to more than ninety per cent. An infinitely dense series of lenses will further increase the probability of transmission to one hundred per cent.

This is an example of the inverse quantum Zeno effect whereby a dense series of measurements of a quantum system along a particular path will force the dynamics of the system to *evolve* along that path. The polarizing lenses can achieve this effect by performing a dense series of measurements along a rotating path of polarization angle. But the inverse quantum Zeno effect is a general phenomenon that can direct any aspect of the dynamics of a quantum system. Imagine an atom sitting at some arbitrary position (A). Position is, of course, a quantum property that may be represented by a wave function or state vector, say $\psi[\{\text{atom at } A\}]$. But like the polarization angle, the position state vector can be represented as a superposition of orthogonal (perpendicular) state vectors that add up to the original state, such as $\psi[\{\text{atom a bit to the left of } A,$ at $B\}$ $(+/-)$ $\{\text{atom a bit to the right of } A,$ at $C\}]$. An equally valid description would be: $\psi[\{\text{atom a bit north of } A,$ at $X\}$ $(+/-)$ $\{\text{atom a bit south of } A,$ at $Y\}]$. In quantum mechanics, each of these representations describes precisely the same state.

We now place a measuring device, such as a powerful electron microscope, over the position B. Since the microscope might detect the atom at B (lying within the atom's position uncertainty), this constitutes a quantum measurement that will project the atom's state vector onto the eigenstates of the microscope. Any atom the microscope detects must be at B, and so will have quantum jumped from A to B. We can repeat this process by placing another microscope at some position D, to the left of B. This measurement will project the atom at B onto its eigenstates. Any atom that is detected will have quantum jumped to D.

The whole process can be continued with a series of microscopes that make quantum measurements along a path $A \rightarrow B \rightarrow C \rightarrow D \rightarrow E \rightarrow F \rightarrow G \rightarrow H$, or indeed along any path we choose. As long as the measurements are sufficiently dense, the inverse quantum Zeno effect will *capture* the atom and drag it along that measured path.

An important point to note about the inverse quantum Zeno effect is that although each measurement along the path may generate a classical signal (as it must to constitute a quantum measurement), the product of the measurement – the measured particle/state – must continue to exist at the quantum level. Only in this way can the next measurement along the line *decompose* the measured state into a new superposition. If, instead, the state that emerges from the measurement is converted to a classical state – perhaps by some kind of amplification process – then the inverse Zeno effect is powerless to modify it. The inverse Zeno effect works only on quantum systems. Classical systems cannot be decomposed into a superposition of states. An irreversible amplification of the quantum system to the classical level represents the end of the line for the inverse quantum Zeno effect.

To get a conceptual feel for what is going on in the inverse Zeno effect, consider again the thought 'note' which, as discussed, can in a sense be *decomposed* into component states: musical note + banknote + written note. Now imagine yourself sitting on a psychiatrist's couch and hearing 'note' and being asked to report the next thought that entered your mind. You might say, 'five-pound' or 'writing pad' or 'piano'. You would be very unlikely (with very low probability) to say the word 'prison'. Now imagine the psychiatrist holding up successive images – a piano, a piano key, a door key, a securely locked door – and then asking you to say the next thing that came into your mind. You still might not say 'prison', but the probability would now be much higher than before. In a sense, the chain of associations have laid down a series of *measurements* in your mind that have taken the thought 'note' to a place where it is much closer to 'prison'. This chain of mental associations has some similarity to how the inverse Zeno effect drags the state vector along a chosen path. Your eventual utterance irreversibly amplifies the thought and becomes the end of the line for the chain of associations; just as an irreversible amplification to the classical level represents the end of the line for the inverse Zeno effect.

The inverse Zeno effect is described as *inverse* because its reverse, the quantum Zeno effect (first described by Misra and Sudarshan of the University of Texas in 1977) describes how continuous measurement can freeze the dynamics of a quantum system. We can see how this works by imagining an atom spontaneously moving along a path $A \rightarrow B \rightarrow C \rightarrow D$. We now fix our microscope securely over A, and perform a dense (in time) series of measurements of the atom at A. However, each time the amplitude of the state vector now drifts off towards B, quantum measurement at A captures the particle and drags it back to A. Continuous measurement at A will prevent the particle moving along the path $A \rightarrow B \rightarrow C \rightarrow D$, thus freezing the dynamics of the system.

It is much easier to perform a dense series of measurements at the same position than to follow a particle along a path; so the quantum Zeno effect has been more intensively studied than its inverse. David Wineland and colleagues at Colorado's National Institute of Standards and Technology succeeded in using the quantum Zeno effect to *freeze* the motion of electrons in a pot of beryllium ions. They first used a laser to *force* the electrons to move from one atomic shell to another, but showed that continuous observation of the electrons would prevent that transition. The quantum Zeno-effect arrests motion by continuous observation; or as John Gribbin describes it in *Schrödinger's Kittens*, 'a watched quantum kettle never boils'.

Both the quantum and the inverse Zeno effect are really aspects of the same phenomenon: the ability of quantum measurement to interact with, and *shape* the dynamics of a system.[9] The special relationship between quantum objects and quantum-measuring devices draws out classical reality from the quantum world. In one last analogy, the process may be compared with the kind of Improvisation Theatre pioneered by Viola Spolin in the 1930s. Spolin's revolutionary approach was to throw away the script. Instead the actors would respond to the audience's reactions and prompting by improvising the ensuing action. At the start of each performance the improvised play can be said to be *indeterminate* just as the word 'note' is indeterminate. The play has certain potentialities dependent on the set of characters present, but no defined plot. With no audience present, we could imagine a *quantum play* in which all possible plots were acted out as a quantum superposition. However, in a real performance, the interaction between the actors and their audience draws out the course of action, the plot for that night's performance. Just as the audience of an improvised play draws out from the infinity of possible plots a single reality for each live performance, so measurement of a quantum system draws out from the quantum superposition of all possible states a single reality for the physical world.

As Niels Bohr said, 'one must never forget that in the drama of existence we are ourselves both actors and spectators'.[10]

But who or what are the actors and spectators inside a living cell of an animal? Is it the cell, the animal, or we who observe the animal? Can living cells draw *out* their own reality or do they need an audience? To answer these questions we next need to explore the nature of physical reality and discover why the world is *real*.

9

What Does It All Mean?

There is a deep mystery at the heart of quantum mechanics. Nobody really understands it. Richard Feynman's advice was to 'not keep saying to yourself, if you can possibly avoid it, "But how can it be like that?" because you will go "down the drain" into a blind alley from which nobody has yet escaped. Nobody knows how it can be like that.'[1] Although perhaps unwise to ignore the advice of one of the twentieth century's most brilliant physicists, I believe we must take a peek down Feynman's blind alley if we are to understand how life works and discover quantum mechanics' limits in living cells.

HOW WONDERFUL IS COPENHAGEN?

The first cogent interpretation of quantum mechanics – the one that has held sway for more than fifty years – is the Copenhagen interpretation. Built by the Danish physicist Niels Bohr, the interpretation also owes much to Max Born and Werner Heisenberg at Germany's University of Göttingen. One of the most revolutionary thinkers of his age, Bohr first realized that the familiar concepts of classical physics must be abandoned in the microscopic realm of subatomic particles. Bohr stressed the importance of epistemology – the study of knowledge, of what we can know about the world. His method was to dissect and examine every statement that claimed to describe either the result of an experiment or reality itself. His scrutiny often revealed ambiguities and unjustified assumptions about the *real world*. He stressed the importance of using words, such as 'wave', 'particle' or 'reality', 'correctly – that is unambiguously and consistently'. In this manner he deconstructed physics at the same time as artists, composers and writers were deconstructing western Art. When in 1917 Marcel Duchamp exhibited a signed men's urinal entitled *The Fountain*, he was questioning the foundations of Art itself – asking what can be called Art? Novels like James Joyce's *Ulysses* were reinventing literature; and Arnold Schöenberg's orchestral works were undermining the tonal foundations of western classical music. At the same time, Niels Bohr was presenting to the scientific world a strange and unfamiliar stripped-down version of physical reality. Nineteenth-century laws everywhere were collapsing before the onslaught of twentieth-century realism. The Age of Reason was dead. It had been trampled into the dirt of Verdun, Arras and the Somme. According to the highly influential physicist John Wheeler, Bohr's job was to, 'build a clean pier for some latter day's bridge to the future'.[2]

Werner Heisenberg gives us a taste of the forcefulness of Bohr's approach:

'For though Bohr was an unusually considerate and obliging person, he was able in such a discussion, which concerned epistemological problems which he considered to be of vital importance, to insist fanatically and with almost terrifying relentlessness on complete clarity in all arguments. He would not give up even after hours of struggling, before Schrödinger had admitted that this interpretation was insufficient, and could not even explain Planck's law. Every attempt from Schrödinger's side to get around this bitter result was slowly refuted point by point in infinitely laborious discussions. It was perhaps from over-exertion that after a few days Schrödinger became ill and had to lie abed as a guest in Bohr's home. Even here it was hard to get Bohr away from Schrödinger's bed and the phrase, "But Schrödinger, you must at least admit that . . ." could be heard again and again.'[3]

Unfortunately, Bohr's insistence on clarity paradoxically endows many of his writings with all the readability of a legal contract. This, for instance, is Bohr on complementarity: 'any given application of classical concepts precludes the simultaneous use of other classical concepts, which in a different connection are equally necessary for the elucidation of the

phenomenon.' Bohr was aware of these difficulties, considering it a fundamental property of language, that there is a 'mutually exclusive relationship that will always exist between the practical use of any word and
attempts at its strict definition.'[4] And also, 'We are suspended in a language in such a way that we cannot say what is up or down.'[5]

The cornerstone of the Copenhagen interpretation is Bohr's statement
that, 'No elementary phenomenon is a phenomenon until it is a registered
(observed) phenomenon.'[6] He insisted that a phenomenon does not exist
until 'closed by an irreversible act of amplification'. The irreversible act of
amplification is the measurement process. You cannot therefore separate
quantum objects from the devices that measure them. Phenomena are
the interaction between quantum objects and measuring devices. No
independent reality is attributed to the quantum objects themselves. There
is simply no such thing as an electron or photon in the absence of
measurement – they do not exist. The fundamental units of our existence
are not atoms, electron or photons but *phenomena*. This, according to
the Copenhagen interpretation, is as far as physics goes. Anything other
than phenomena is philosophy or theology. According to Wheeler, Bohr
'did not propose an answer, not philosophise, not go an inch beyond the
fullest statement of the inescapable lessons of quantum mechanics.'[7]

But where within a phenomenon does the quantum system stop and
the measuring device start? In the Copenhagen interpretation of the
two-slit experiment, each photon or electron travelling through the slit
screen must be described by the wave function or probability wave until
the act of measurement. The electron is initially considered to exist as
a quantum superposition of position states before it couples with a
measuring device placed over one slit. In the Copenhagen interpretation,
measuring devices are always considered as classical objects. The coupling
between the measuring device and the electron superposition is proposed
to 'collapse the wave function', so that the electron plus measuring device
collapses into a single classical reality – the electron detected in one or
other of the slits. The difficulty arises when you ask, at what stage during
the interaction between the quantum system (the electron) and the measuring device does the collapse of the wave function take place? The
measuring device might measure the disturbance to an electromagnetic
field consequent on the passage of a changed electron. But an electromagnetic field is made of photons that can certainly exist in the same kind

of superposition of states which the electron inhabited prior to its measurement. Changes in the electromagnetic field might then be transmitted to components of the measuring device which would, after amplification, result in an audible signal. But the atoms and molecules of the measuring device are also composed of the same kind of fundamental particles (protons, electrons etc.) that can be made to demonstrate quantum superposition phenomena. Shouldn't the measuring device also be able to exist in a superposition of states?

We never see measuring devices in a superposition of states – clicked and not clicked – so somewhere between the original quantum event and our observation of an audible signal, the quantum wave function must collapse to make a classical reality. But when? Bohr's attitude was that it doesn't really matter where the line is drawn: 'the place within each measuring procedure where this discrimination is made is . . . largely a matter of convenience.'[8] He compared the problem to the difficulty in 'placing of the separation line between object and subject in the analysis of various aspects of psychical experience.'

THE CAT AMONGST COPENHAGEN'S PIGEONS

Bohr's colleague, Erwin Schrödinger was not convinced and devised a thought experiment (often called a *gadanken* experiment) to make his point. The now famous Schrödinger's cat paradox has fascinated scientists and philosophers ever since. Schrödinger questioned the use of the wave function to describe reality and pointed out that, 'one can even set up quite ridiculous cases. A cat is penned up in a steel chamber along with the following diabolical device . . .'[9] Schrödinger's diabolical device was a Geiger counter and pot of highly radioactive material.

However, to remain on familiar ground, we will utilize our experiment with two holes to perform essentially the same function as Schrödinger's device. Imagine our two-slit experiment placed entirely within a sealed box alongside the story's eponymous feline. We place a detection device (ignoring its workings for the moment) capable of registering the passage of single photons at the right slit. When the detector records the passage of a photon passing through the right slit, it will send a small signal to a dastardly contraption that releases a vial of poisonous gas into the box, thereby killing our (imaginary) cat.

We fire a single photon through the screen. The photon has a fifty per cent chance of travelling through the left slit, in which case the detector will not fire and our famous cat will live out the rest of her days comfortably curled up at Schrödinger's fireside. However, if the photon travels through the right slit, the detector fires and it's curtains for our feline friend. But as we have painstakingly demonstrated, in this kind of experimental set-up, single photons pass through both slits simultaneously as a quantum superposition. But there is nothing in quantum mechanics that states that the superposition must stop with the photon. The detector will become *entangled* in this superposition and must also exist as a superposition of a detector that does not fire (because the photon went through the left slit) and a detector that fires on detecting the passage of a photon through the right slit. But quantum mechanics demands more. The photon-detector is equally entangled with the poison vial which must also exist in a quantum superposition of intact and broken (complete with escaped poison). The photon plus photon-detector plus poison is just as surely entangled with the fate of our cat. The superposition of the photon/photon-detector/poison leads irrevocably to a quantum superposition of both a live and dead cat. The photon's capacity to exist in two places at once – at the right and left slits – has become entangled with that of our cat who must now exist in two states at once: live and dead.

Much ink has been spilt pondering the fate of Schrödinger's cat. In *A Brief History of Time*, physicist Stephen Hawkings claims whenever he hears mention of *that cat*, he reaches for his gun. Yet, despite Hawkings' animosity, the fate of Schrödinger's cat neatly illustrates the problem of Bohr's dichotomy between the quantum world and everyday reality. What precisely is the measuring device in this scenario? Is it the photon-detector? Bohr's argument that measuring devices must be classical was based on the premise that their result must be communicated to us as classical observers. However, with the box closed, the result of the photon-detector's measurement (plus or minus photon *or* plus and minus photon) is communicated only to the dastardly device, so there is no requirement for it to enter the classical world of our perceptions. Is it then within the device and its poison that the classical world intervenes?

But its result is communicated only to our unfortunate cat. Must our cat obey classical laws? Although we may not be able to see, hear or talk in quantum terms, who is to say that a cat can't? Should we impose the limits of our perception on the cat? Perhaps it is quite happy to exist in a state that is a quantum superposition of being dead and alive? It might be better than just being dead. If not the cat, then it must be when a conscious human observer – with the same limitations as ourselves – opens the box. The quantum superposition of photon/detector/dastardly device/poison/dead-and-alive cat must then surely collapse to become a classical reality.

But can the state of a photon really depend on whether or not a conscious human observer is in the room to watch it? Many supporters of the Copenhagen interpretation take the view that it does. Eugene Wigner imagined a person inside the box instead of the cat: Wigner's friend, as he came to be known. The Copenhagen interpretation of quantum mechanics would describe Wigner's friend by a wave function that would become entangled with the photon in the same irrevocable way as Schrödinger's cat. But, if on opening the box and finding his friend alive, Wigner were to ask what kind of state he was in before the box's opening, the friend would reply that he experienced nothing unusual. He would have no memory of being in a superposition of both dead and alive. Wigner concluded that 'a being with consciousness must have a different role in quantum mechanics than the inanimate measuring device . . .'[10]

This requirement for the presence of conscious observers seems to define a world with ourselves acting as arbiters of reality. Without us, photons, electrons, protons and bigger objects such as measuring devices, living cells – and possibly even cats – are supposed either not to exist at all, or as some kind of fuzzy quantum potentiality. Quantum-measuring devices are proposed to be endowed with the ability to translate quantum events into classical reality, but only when we are there to act as observers. But who is to say that a cat cannot collapse the wave function? Indeed, perhaps cats do it all the time and we owe our own classical reality to their unstinting observations of *our* activities. If we accept cats as sufficiently conscious to perform quantum measurement, then how about the frog physicist in the last chapter? Can an amphibian collapse quantum wave functions? What about an ant or even a bacterium? Surely bacteria cannot be conscious, but a bacterial cell could easily be placed in the same position as Schrödinger's cat. Would the bacterium exist as a quantum superposition of a cell dead and alive until we came along to collapse its wave function? As Chapter Twelve will show, consciousness is a very slippery concept on which to pin our world's reality.

Another problem with a consciousness-dependent version of reality is that, as far as we know, consciousness is a relatively recent invention. How real was the universe before consciousness evolved? The physicist John Wheeler has taken the consciousness-dependent reality view to its logical conclusion, proposing that we live in a 'participatory universe', wherein the universe depends for its existence on conscious observers to make it real, not only today but retrospectively right back to the Big Bang! Wheeler suggests that the presence of observers imparts a 'tangible "reality" to the universe, not only now but back to the beginning',[11] by a kind of backward-acting wave function collapse. In this scenario, the universe existed in an undetermined ghost state until the first conscious being opened its eyes to collapse the wave function for the entire universe and bring into being its entire history, including the geological and fossil record recording its own evolution.

This consciousness-participatory approach is perilously close to Descartes' solipsism '*Cogito ergo sum*' – I think therefore I am; or Bishop Berkeley's '*Esse est percipi*,' – to be is to be perceived. Biologists, including myself, are profoundly uneasy with the concept that the objects of their study became *real* – in geological time – only a few moments before. The consciousness-participatory interpretation of quantum mechanics seems to allow the human psyche to play a pivotal role in defining the external world. Most scientists are very reluctant to reverse the triumphs of Copernicus, Galileo and Newton and place man, once again, in the centre of the universe. To the concept that the world does not exist outside human perception, most people would, I believe, adopt a similar attitude as that of Dr Johnson, as described in Boswell's *Life of Johnson*. Boswell relates how he and Johnson were discussing Bishop Berkeley's theory of the non-existence of the material world. Boswell remarked that though no one believed the theory, it could not be refuted. Johnson kicked a large rock and replied, 'I refute it thus.'

Einstein was also deeply unhappy with the participatory reality described by the Copenhagen interpretation. He claimed, 'Out yonder there was this huge world, which exists independently of us human beings.'[12] In a famous paper written by Einstein, Boris Podolsky and Nathan Rosen in 1935, Einstein challenged Bohr's viewpoint. The paper was only the latest in a series of challenges to the Copenhagen School mounted by Einstein between 1920 and 1940. Despite their differences, both of their writings displayed deep mutual respect and good-natured affection. For instance, from Einstein: 'I am studying your great works and – when I get stuck anywhere – now have the pleasure of seeing your friendly face before me smiling and explaining.'[13] Or from Bohr: 'For such endeavours of seeking the proper balance between seriousness and humour, Einstein's own personality stands as a great example.'[14] Many of their exchanges took place at various conferences. Heisenberg describes the general character of these meetings:

'The Solvay conference in Brussels in the autumn of 1927 closed this marvellous period in the history of atomic theory. Planck, Einstein, Lorentz, Bohr, de Broglie, Born and Schrödinger, [were there . . . but] the discussions were soon focused to a duel between Einstein and Bohr . . . We generally met at breakfast at the hotel, and Einstein began to describe an ideal experiment in which he thought the inner contradictions of the Copenhagan interpretation particularly visible . . . During the meeting and particularly in the pauses we younger people, mostly Pauli and I, tried to analyse Einstein's experiment, and at lunch time the discussion continued between Bohr and the others from Copenhagen. Bohr had usually finished the complete analysis of the ideal experiment late in the afternoon and showed it to Einstein at the supper table. Einstein had no good objection to this analysis, but in his heart he was not convinced.'[15]

PREDICTING REALITY

Paradoxically, Einstein's challenges led to a strengthening of the Copenhagen interpretation's position. Each of his objections was so effectively rebutted by Niels Bohr and colleagues that the theory emerged the stronger. His 1935 paper, which became known as the EPR paper, was Bohr's toughest test. The paper begins with the statement; 'A sufficient condition for the reality of a physical quantity is the possibility of predicting it with certainty, without disturbing the system.'[16] Einstein goes on to describe a thought experiment in which two quantum systems interact. We can imagine one particle with a momentum of, say, ten arbitrary units decaying into two particles, *p1* and *p2*. The two particles then speed off in opposite directions. Since the system's total momentum was known before the collision, then from the law of conservation of momentum, the total momentum of the pair is also known after the collision, for it must equal the same ten units. What is not known is how that momentum is shared between the two particles. In the standard Copenhagen interpretation, the two particles are said to be *entangled* in what has become known as an *EPR pair*. The particles do not have an independent reality but are described by a single wave function, containing within it all of the possible states of both particles (for example: ψ\{p1 with momentum of 1 unit and *p2* with momentum of 9 units\} $(+/-)$ \{*p1* with momentum of 2 units and *p2* with momentum of 8 units\} $(+/-)$ \{*p1* with momentum of 3 units and *p2* with momentum of 7 units\} ... \].

In the standard Copenhagen interpretation, the momentum of a single particle has no meaning until a measurement is made. The property of a single particle's momentum simply does not exist before measurement. Einstein's challenge to this position was to point out that it is unnecessary to measure both particles to gain definite information about both. Since the total momentum of the entangled pair is known, it is possible to measure the momentum of one and thereby predict, 'with certainty, without disturbing the system', the momentum of the other particle. If the momentum of a particle can be predicted with certainty without disturbing it, then it must possess an *objective reality* in the absence of measurement. If that was true, then the Copenhagen interpretation of quantum mechanics was wrong.

According to one of Bohr's colleagues, Rosenfeld: 'This onslaught came down upon us as a bolt from the blue. Its effect on Bohr was remarkable ... Everything else was abandoned: we had to clear up such a misunderstanding at once.' The answer was, however, more elusive than first imagined and with 'growing wonder at the unexpected subtlety of the argument',[17] Bohr and his colleagues set about formulating a reply to Einstein's paper.

Bohr's eventual response was to question Einstein's restriction, 'without disturbing the system'. In typical prose, he argued that the 'procedure of measurement has an essential influence on the conditions on which the very definitions of the physical quantities in questions rests.'[18] This rather inscrutable sentence and the paper that followed it, insist that the entire experimental set-up constituted a disturbance to the system. In particular, neither the measured nor unmeasured particle can be considered to have an objective reality in the absence of measurement by the entire experimental set-up. Contrary to Einstein's contention that measurement of the second particle does not disturb it, the measurement disturbs the particle in a very fundamental way – by making it real. No matter that the entangled pair of particles are apparently separated in space, because prior to measurement they are actually not separated at all – they are not independent. Measurement of one of the EPR pair instantaneously makes the other *real*, irrespective of their eventual degree of separation.

Bohr won the argument, but the price of his victory was extraordinary – nothing less than an abandonment of our notion of objective reality, that there is a world out there independent of our experiences. Although the EPR paradox was developed with only a single property – momentum – the argument can be applied equally to any property of matter or radiation such as position, energy, charge, spin, polarization state – none is considered to have objective reality in the absence of measurement. But then what are matter and energy and our whole existence other than a collection of properties like momentum, energy and charge? For living systems the problem is particularly acute, since the Copenhagen interpretation seems to leave them in a kind of quantum limbo until measurement is made; presumably when an intelligent observer arrives to measure their state (although in the strict interpretation, who or what is making the measurement is not defined). Can we really believe that bacteria do not exist until they make us sick? Did the entire biosphere exist as a massive superposition of every possible state until humanity evolved?

Einstein accepted the logic of Bohr's position but concluded that it was 'so very contrary to my scientific instinct that I cannot forgo the search for a more complete description.'[19] In fact, Einstein was to spend much of his remaining scientific career in a fruitless search for that 'more complete description'.

THE PROOF OF THE PUDDING

But was Bohr right? Is independent objective reality an illusion? When Einstein and Bohr debated the EPR paradox, it existed only as a thought experiment. Nobody thought that EPR entanglement would lead to testable predictions any different from those generated by classical entanglement. To illustrate this, consider again our example of two EPR-entangled particles with a total momentum of ten units. Suppose a scientist (Alice) performs a measurement on the first and finds it has a momentum of seven units. Classical conservation laws are now all she needs to predict that the second particle must be left with a momentum of three units. She could ask her colleague (Bob) to perform the measurement of the second particle (which might be one mile away by now) and perhaps inform her of his result – by phone. A dutiful scientist, Alice would not rely on a single observation but would perform a series of measurements on many EPR-entangled particle pairs. In each case, she would measure the momentum of the first particle, simply subtracting its momentum from ten, to predict the result of Bob's measurements. At the end of her experiments she would ask an independent colleague (Carol) to compare her predictions to Bob's measurements. So long as no errors had slipped in, Carol would discover that Bob's measurements exactly matched Alice's predictions.

But this result would not distinguish between the classical and the quantum version of events. Both theories make the same prediction about the expected correlations between Bob and Alice's measurements: they should always add up to ten. The theories differ only in their model for the state of both particles prior to measurement. Classical theory predicts that the momentum of the particles was fixed at the moment the original single particle decayed into two particles. Quantum theory insists that the property of momentum for either entangled particle did not exist until one was measured.

In 1962 Irish mathematician John Bell discovered that under certain circumstances classical and quantum theory would make different predictions for the pattern of correlations expected for EPR-entangled particles. In particular, if Bohr was right, the EPR correlations should violate a mathematical relationship that has since become known as Bell's inequality. The correlations are of a subtle nature and in themselves do not disprove a classical version of reality. Many phenomena may be correlated. The weather in London is pretty well correlated with the weather in Guildford. The cities are close enough (roughly 30 kilometres apart) to suffer the same rainy weather systems. If a colleague working in London were to keep a record of which days it rained outside her office, the records would pretty well match a similar record I might make in Guildford. The correlations are due to an atmospheric communication between the two cities. It would be considerably more surprising if I found the same pattern of correlations for rain records kept by another colleague in San Francisco. To account for the phenomenon, I might propose some kind of ultra-fast signalling of weather conditions between London and San Francisco. Moreover, if the onset of rain was at precisely the same (GMT) time in the two cities then classical theory (particularly relativity) would run into problems, since it forbids instantaneous communication. Nevertheless, John Bell demonstrated that if quantum mechanics was right, then measurement of EPR pairs of particles should demonstrate correlations which would be instantaneously communicated to all parts of the quantum system, *as if space did not exist between them.*

A number of experimental tests of Bell's inequality have since been performed. The most famous was made by Alain Aspect and colleagues at the University of Paris-South. Aspect measured polarization state of EPR-entangled pairs of photons that were emitted from atoms of calcium. Each pair was emitted from a single atom in such a way that their polarization states must be at right angles (if one of the photons was found to be polarized at say twelve o'clock then its partner had to be polarized at three o'clock). But the Copenhagen interpretation of quantum mechanics claims that, prior to measurement a polarization state does not exist as a real property. Yet, as we discovered in the last chapter, we can influence the polarization state of a photon by measuring it. If we turn a Polaroid lens to two o'clock, all the photons that emerge from the lens will be polarized at two o'clock. But if the photon we measure is an EPR-entangled particle, we would also know that the partner to that photon must be polarized at eleven o'clock (perpendicular to two o'clock). If quantum mechanics is right, the polarization angle we choose to measure fixes, not only the polarization state of the particle we are measuring (which is odd enough) but also the polarization state of its (by now distant or *non-local*) partner. It is this non-classical influence of our choice of measurement on the state of the distant particle which leads to the violation of Bell's inequality.

Aspect arranged for EPR-entangled photon pairs to be fired at a pair of detectors. A random switch, operating at an extraordinary ten thousand million times a second, changed the polarization angle to be measured for each detector. The rapid switching ensured that no signal travelling at the speed of light could pass from one detector to the other (about ten metres away) in the time it took a photon to reach the detector. So long as signals could not travel faster than light, then one detector could not *know* the angle the other was measuring. The classical version of events would insist that in these circumstances the particles have independent reality so it should not matter what measurements happened to be performed on one particle: those choices should not affect the properties of its partner. Yet they do. When the measurements were compared, they demonstrated the type of correlations which violated Bell's inequality. The properties of each photon depended, not only on the measurement performed upon it but on the measurements performed on its partner.

The Aspect experiment demonstrated that non-local connections exist between two objects (photons) separated by ten metres. More recent experiments have confirmed and extended Aspect's work – and non-locality has now been demonstrated for EPR particles separated by more than a kilometre. EPR entanglement is real enough to have practical applications. It is currently being investigated as a way of transmitting uncrackable coded information. Particles have even been *teleported* from one place to another using EPR entanglement. But the phenomenon has even greater significance for our view of reality. Remember that EPR entanglement, as originally envisaged by Einstein, requires only that the particles have interacted some time in the past. But if we go back far enough, right back to the Big Bang, then all particles in the universe have interacted. Every particle becomes connected to every other particle in a single massively entangled super-EPR quantum state. This state will persist until measured. The Copenhagen interpretation is then stark – the world does not exist until we (whoever *we* are – that is not made clear) measure it.

When we make an observation, we make the world real and it becomes real everywhere instantly. If you find that explanation hard to swallow, the escape hatch is to allow signals to travel faster than the speed of light. Such signals could communicate information (in the Aspect experiment, the polarization angle to be measured) between the detectors to establish the EPR-type correlations between particle measurements. But remember that ERP-linked particles may find themselves anywhere in the universe. Can a photon here on Earth be in instantaneous communication with an EPR-paired photon in the Andromeda galaxy? Such signals would wreck havoc with causality and create big problems for that other great theory of twentieth-century physics: relativity. Einstein did not live to see the Aspect experiment so we can only speculate as to which aspect of his worldview he would have preferred to relinquish: objective reality or relativity theory. Or perhaps his genius would have found another way out of the impasse – maybe the right way.

LIFE AFTER COPENHAGEN

The EPR paper, Bohr's response *and* Schrödinger's cat paradox paper were published in 1935. A great debate involving all the combatants should have followed. But those were dark days in Europe and the debate's protagonists were soon scattered by the catastrophe of the Second World War. Max Born built much of the mathematical formalism of quantum mechanics. It was he who first proposed that the wave function of a particle should represent a probability wave. Appointed Professor of Physics in Göttingen in 1921, Born made its university one of the chief centres for particle physics research in the world. But, as a Jew, he was forced to flee Germany in 1933, settling in Edinburgh where he remained until his retirement in 1953. Einstein, also Jewish, was appointed as Associate Professor at Zürich University in 1909 but, facing the rising tide of Nazism, he fled to America, also in 1933. Once settled in Princeton, his thoughts turned increasingly to the dangers of nuclear technology. In 1939 he wrote to President Roosevelt to warn him of the possible threat of a Nazi atom bomb. Experiments performed by Otto Frisch in Niels Bohr's laboratory had demonstrated the feasibility of nuclear fission, alerting Einstein (and other scientists) to the possibility of a bomb. The Allies, fearful of Bohr's expertise falling into Nazi hands, sent Bohr the offer of a job in England – with the message printed on a microfilm disk drilled into the tooth of a courier. After much deliberation, Bohr fled Denmark and his beloved Institute, in a rowing boat – spirited away by British Intelligence, with the Nazis at his heels.

The German project to develop a nuclear bomb was led by Bohr's former colleague Werner Heisenberg. Fortunately for us, although Heisenberg was a brilliant theoretician, he was a less successful practical scientist. His fission research project never came close to building a bomb. Erwin Schrödinger occupied the chair of theoretical physics at Berlin. Though Catholic, he decided he could no longer live in a country in which persecution of Jews had become a national policy. In 1933, he received an invitation from Eamon de Valera to take up the Chair of Theoretical Physics at the newly founded Dublin Institute for Advanced Studies. Schrödinger settled in Dublin until his retirement in 1956. The founder of quantum mechanics, Max Planck lived out the war in Germany but when in 1944 his only surviving child, his second son Erwin (his other son was killed in the war and both daughters died in childbirth) was executed for involvement in a plot to assassinate Hitler, Planck was devastated and died broken-hearted a few years later.

With the architects of quantum theory scattered by the chill winds of the Second World War, any progress in its interpretation came to a virtual halt. Sir Rudolf Pierls, who held the Chair of Physics at Oxford University until 1974, claimed that 'There is only one way in which you can understand quantum mechanics . . . the Copenhagen interpretation.'[20] Yet dissatisfaction with quantum mechanics according to Bohr has grown over recent decades. The question of what constitutes a quantum measurement was never fully addressed by Bohr and has increasingly come under scrutiny by new experimental approaches which probe the border between quantum and classical behaviour. Several new interpretations are now on offer. They all make precisely the same predictions as the standard Copenhagen interpretation, yet offer different insights into the quantum world. We will explore just a few.

THE FAIRIES

The development of the atomic bomb led to a continuing, inevitable politicization of particle physics. Having worked briefly on the American atom bomb project, after the war David Bohm was called to testify before the Un-American Activities Committee. Refusing to give evidence against his former colleagues, he was indicted on charges of contempt of Congress. Though acquitted, he fell victim to the anti-Communist McCarthyite witch-hunts and found it impossible to obtain a post in the United States. He left to take up a post at London's Birkbeck College, where he developed the *hidden variables* interpretation of quantum mechanics.

The hidden variables approach considers that properties such as position, momentum and energy possess *objective reality* at all times, irrespective of measurement. Heisenberg's uncertainty principle is regarded as a fundamental limitation of our powers to measure these properties to arbitrary degrees of accuracy, not as a limitation on the reality of those properties. The actual value of properties, such as momentum, position or polarization angle, is determined by properties (variables) we are unable to measure. Since we cannot determine the value of these variables, they are forever hidden from us (hence hidden variables). Objective reality is restored, though it cannot be entirely known.

In Bohm's theory, causality is also restored. Quantum randomness (for instance, the impossibility of predicting at which slit a photon may be detected in the two-slit experiment) becomes pseudorandomness due to incomplete knowledge. It is the hidden variable (for instance, the precise value of a photon's momentum) that determines which way a photon will travel. Similarly, hidden variables govern when a radioactive atom will decay, or whether a photon will pass through a polarizing filter.

Until fairly recently, a belief in hidden variables was considered the physicist's equivalent of believing in fairies. Hidden variables are, at all times and in all possible situations, hidden. Why postulate the existence of things that can never be seen? You may as well postulate that fairies determine when a radioactive atom will decay or through which hole a photon will pass. (The reply of my Bohmian colleague Jim Al-Khalili is that it is better to believe that fairies are the cause of strange goings on

at the bottom of your garden than to deny the existence of your garden)
However, an additional problem for the hidden variables approach is to
account for the EPR-like correlations that arise from measurement of
particles (the Aspect experiment or indeed the frog physicist version of
the two-slit experiment described in the last chapter). Bohm borrowed a
concept termed a *pilot wave* from a pioneer of quantum mechanics, Louis
DeBroglie. Particles are considered to be conventional particles at all
times, but their trajectory is governed by a pilot wave which steers the
particle along its course. Bohr insisted that any description of a quantum
system must include the entire experimental set-up. But how do single
particles *sense* the entire experimental set-up? In the hidden variables
approach, it is the pilot wave that does this job. The pilot wave behaves
like the wave function of the Copenhagen interpretation to travel simultaneously through both slits of the two-slit experiment. It is the pilot
wave which *senses* the presence of a detector over one slit and communicates that information instantly to all particles, causing small shifts in
their trajectory, thereby destroying the interference pattern.

The problem with the pilot wave is that it must possess such peculiar
properties. It must transmit information between all parts of a quantum
system, instantly; its strength should not reduce with distance between
objects. Indeed Bohm has proposed that the universe is filled with an
infinite number of overlapping pilot waves, connecting all matter in an
undivided whole. Yet the pilot wave should not carry the kind of information that could be used to send a classical signal. Prod one particle
on Earth and its EPR-entangled partner on Alpha Centauri should not
jump. Yet it must transmit information about an entire experimental
set-up between all particles involved. How does the pilot wave *know* what
information it is allowed to transmit? In the two-slit experiment, it must
be able to *see* a detector placed above one slit when turned on but not
when turned off. What decides the level of interaction that constitutes a
measurement which must be communicated through the pilot wave? As
you can see, the measurement problem still remains a problem in the
hidden variable approach.

In a sense, the hidden variable theory does not do away with the
peculiarities of quantum mechanics: it packages them differently. While
the Copenhagen interpretation of the quantum wave function is a combination of a strange wave and an odd particle, the hidden variable approach

puts all the weirdness into the pilot wave so that wave function becomes a combination of a sensible particle and a strange and odd wave. To a certain extent, classical reality is restored but at a cost. We must accept the physical reality of a pilot wave with properties unlike anything else known to physics.

Despite these difficulties, the hidden variables approach works, making the same predictions as the standard Copenhagen interpretation for all experiments ever performed. There is no experimental data to choose between the rival interpretations. The pilot wave concept may appear an inelegant artifice in comparison to the austere beauty of the Copenhagen interpretation, yet that seems a small price to pay for a system that restores both objective reality and causality to our world.

MY GOD, THERE'S TWO OF HER!

A favourite theme of 1940s Hollywood films was women who had doubles – often a twin who would represent the character's dark side. In *Cobra Woman* (1944), Maria Montez plays both an innocent island girl and her alter ego, a seductive snake-worshipping high priestess who tosses her victims into a smoking volcano. Inevitably, these two aspects of womanhood would compete for the heart of some hapless male who would cry 'My God, there's two of her!' at the film's moment of revelation. Fantastical though such genre movies were, they pale into commonplace beside the Many Worlds interpretation of quantum mechanics. In this interpretation of reality, we do not have a single double but perhaps an infinity of doubles, each occupying their own parallel universe.

The Many Worlds interpretation takes quantum mechanics at face value. In the two-slit experiment, when a single photon is fired through the screen, the photon behaves as if experiencing interference. The standard Copenhagen interpretation claims that the photon travels through both slits simultaneously as a quantum superposition and thereby generates interference. The hidden variables approach envisages a pilot wave travelling through both slits and adjusting the trajectory of the single photon that travels through only one slit. The starting point for the Many Worlds approach is the observation that the trajectory of a single photon is modified, *as if* it were experiencing interference, with another photon taking the alternative route. In 1957 Hugh Everett III, a graduate student in John Wheeler's Princeton laboratory, proposed a simple – but revolutionary – modification to the above statement – simply drop the *as if*. Instead, it is envisaged that there are actually two photons which interfere with each other after travelling through both slits. But then why do we see only a single silver grain when the two photons land on a photographic plate? Why do we only hear a single click when the two photons pass through both slits, each armed with a detector? Because the other photon is in another universe.

In Everett's approach, the universe splits every time a quantum system is faced with a choice. If there are two possible paths for a photon, then two universes will exist side by side. In one universe the photon will take one path (for instance the left slit in the two-slit experiment). In the other universe the photon will take the other path (the right slit). Initially the alternative universes exist in the same space, allowing the two photons to generate the interference pattern. But once a measurement is made to determine the path of a single photon, then the two universes become independent. In one universe the photon will have passed through the right slit. In the other it will have passed through the left. The alternative universes are now fully separated, do not interfere with each other and consequently do not generate an interference pattern. But it cannot be just the photon that is different in the two universes. In one universe there will be an experimenter who will hear a click coming from the left detector; and in the other universe there will be a *twin* experimenter who will hear a click coming from the right detector. So we have two parallel universes existing side by side but with observers in each witnessing alternative versions of reality.

Yet we have considered only the trajectory of a single photon. There are billions of quantum events taking place within each ounce of matter every second. Each event may have thousands of alternative outcomes (consider the trajectory of a photon in the absence of the slit screen). Each possible outcome must be associated with a universe in which that outcome is realized. A multitude of these parallel universes then make up the *quantum multiverse* which contains every possible alternative version of reality.

The advantage of the Many Worlds interpretation is that it makes very few assumptions. There is neither an *ad hoc* collapse of the wave function nor any nebulous pilot wave. The standard equations of quantum mechanics describe a quantum system in a superposition of states; the Many Worlds interpretation proposes that this superposition is the multiverse. Measurement acts as a kind of filter that allows through into our world only one of the possibilities existing within the multiverse.

To get a feel for the extraordinary implications of the Many Worlds hypothesis, imagine a comet or meteorite travelling through our solar system some sixty-five million years ago. If we travel further back in time, then we will certainly encounter billions of tiny quantum events that might have influenced the precise trajectory of this meteorite. Perhaps, millions of years earlier, the disintegration of a radioactive atom on its surface led to the ejection of a heavy particle from the meteorite and a consequent tiny shift in the orbit of this lump of rock. Millions of years later that tiny shift has been magnified by a billion tiny interactions and is now responsible for a significant modification to the trajectory of the meteorite, so that it passes close by but just misses a planet, the Earth. The dinosaurs go about their business with only an upward glance at the bright object hurtling through the heavens. With no dinosaur extinction, mammals remain a relatively insignificant group of egg-stealers. Perhaps the dinosaurs then evolve further to become intelligent, develop civilization and maybe one of them is at this very moment writing a book about quantum evolution.

The point is that, however fanciful this scenario seems, in the Many Worlds hypothesis it is almost certainly happening in one of the many trillions of possible universes, existing out there in the multiverse. The precise course of the meteorite that led to the dinosaurs' extinction was determined by billions of quantum events that, according to the Many Worlds hypothesis each would have led to a splitting of the universe. At every juncture where one event happened in our universe an alternative event occurred in a parallel universe. In trillions of the alternative universes, the meteorite that wiped out the dinosaurs would have missed. We can only guess what the evolutionary pathways of both dinosaurs and mammals would be in the absence of the Cretaceous-Tertiary extinction event, but it would certainly be (or is) a very different world from the world we inhabit.

For every event in history there must be billions of alternative scenarios, each separated from our world by a different sequence of quantum events. You can invent your own alternative universe scenario. Perhaps an iceberg broke away from the polar ice sheet in 1910, drifting through the North Atlantic for two years until the night of the 14th April 1912, when it narrowly missed a great ocean liner. The passengers were happily unaware of their close encounter with tragedy and the fact that they had been saved by a molecular vibration which had, months before, dislodged a chunk of ice from the iceberg. The loss of that lump had shifted the course of the iceberg just enough to cause it to drift harmlessly past the liner (a quantum version of the well-known butterfly effect whereby small events may have big repercussions). The passengers reach their destination and lead full and active lives, marrying, having children and making their world significantly different from ours.

For every significant – or indeed insignificant – event in the world's history, it is always possible to find some critical quantum event that would change that history. If we consider history all the way back to the Big Bang, then the choice of alternative histories is immense and the Many Worlds hypothesis predicts that there are an astronomical, probably infinite number of universes. In the Many Worlds *multiverse*, anything that can happen will happen. The physicist Paul Davies commented that the Many Worlds interpretation is 'cheap on assumptions but expensive on universes.'[21]

Many Worlds has many adherents, particularly amongst cosmologists, since, in contrast to the Copenhagen interpretation, it does not require any division of a quantum system into the measured and an external-measuring device. Cosmologists need to treat the entire universe as a quantum system and therefore do not have recourse to any external measuring device (except perhaps a deity) so the Many Worlds approach tends to be their favourite. But for inhabitants of the universe there remains a problem. Why do we see interference effects in some situations but not others? In the two-slit experiment when we don't detect individual trajectories, we get interference patterns indicating that the parallel universes have not separated, but occupy the same space – a double universe. If we place a detector over one slit, the interference patterns disappear – the universe has split into two independent universes. What decides the point at which the double universe splits to become two independent universes? Presumably this is the point at which a quantum measurement is considered to have taken place. But this brings us back to the old problem of what constitutes a quantum measurement. Does the living cell inhabit the multiverse or a single universe? We could cannibalize the Wigner approach and propose that the universe splits whenever the result of the measurement is recorded in the mind of a conscious observer but this is precisely the kind of metaphysical hot water we hoped to avoid when looking for alternatives to the Copenhagen interpretation. Once again, the measurement problem remains a problem in the Many Worlds interpretation.

TIME TRAVEL

Aspect's experiment covered earlier indicates that the world cannot be both local (objects affected only by their local causally connected environment) and real, inbetween measurements. In the Copenhagen interpretation it is reality that is sacrificed. The Many Worlds approach is an inverse of the Copenhagen interpretation – everything is real until measurement. Yet it achieves the same end, since a *single reality* does not exist until after measurement. The Hidden Variables approach sacrifices locality, with all particles in the universe being connected in a kind of holistic unity, until measurement.

The *transactional interpretation* of quantum mechanics similarly sacrifices the principle of locality by allowing signals to travel backwards in time. This approach grew out of work by Richard Feynman and John Wheeler who suggested that electromagnetic waves may travel both forward and backward in time. John Cranmer of the University of Washington in Seattle extended this suggestion to propose that EPR correlations are established by particles signalling backward-in-time. In, for instance, the Aspect experiment, the measuring apparatus that absorbs one of the photons in the EPR-entangled pair is proposed to send a signal (a transaction) that travels backward-in-time to the particle that emitted both photons. This transactional signal *tells* the particle to emit the paired photons with the correct polarization to match the angle measured by the filters.

Cramer and other adherents to the transactional interpretation are careful to qualify the nature of the backward-in-time transaction to exclude it from transmitting a classical signal. There is (as yet) no means of sending a signal to inform your younger self of the winner of last year's Derby. Yet, real backward-in-time signalling, with its paradoxes and problems with causality, cannot be entirely ruled out in the transactional interpretation, which makes many physicists deeply sceptical about the approach. From our point of view, it, like the other interpretations, fails to address the measurement problem we are interested in. It does not tell us the limits of backward-in-time signalling within a living cell. Can a protein, for instance, signal backward-in-time to the DNA that encoded it? The transactional interpretation does not provide an answer to the measurement problem.

Backward-in-time signalling, the universe as a holistic whole, an infinity of universes or an unreal world? Which do you prefer? If these don't suit, there are about a dozen other interpretations to choose from. Yet beware. None restores a classical version of reality. Each takes the strangeness of quantum mechanics and packages it in some way. But the strangeness remains, whatever the packaging.

The question we initially posed remains unanswered. Where, within a living cell, does a quantum measurement take place? Fortunately a relatively recent approach, independent of the interpretations of quantum mechanics, gives us a way forward, though probably not the entire answer.

DECOHERENCE

Consider a single photon in a closed box. If we have no knowledge about this photon's polarization state then it may subsequently be found with any angle of polarization, →, ↖, ↗ or ↑ or any angle inbetween. In quantum mechanics we describe the state of this photon with a wave function made up of a superposition of all these possible states: $\psi[\{\rightarrow\}$ $(+/-) \{\nwarrow\} (+/-) \{\nearrow\} (+/-) \{\uparrow\}]$. Our classical minds might ask, 'but which state is the photon actually in?' However, if we are only considering the photon in isolation, then these states are all the same, even classically. If there is no external reference direction then →, ↖, ↗ and ↑ are indistinguishable – they are the same state. It is only by placing the photon in an environment with reference points on which one can map out particular directions, that the state ➔ can become distinguishable from ↖. It is only after this mapping takes place that it makes any sense to talk of a photon having a direction of polarization at all.

Similarly, the concept of position makes sense only when a particle is mapped to a space. In empty space without external reference points, position loses its meaning. The same considerations apply to concepts like energy, momentum or time – they can only be defined with reference to externally mapped reference points and/or an external clock. So, in order to define a particular state it is necessary to have an environment and only by reference to this environment do classical concepts like a single defined direction of polarization, or a single energy state of a single position, become real. This is one aspect of any measuring device; it allows the quantum system to be mapped to an external environment, and thereby legitimizes classical concepts such as position or momentum. In the absence of an external environment, the quantum wave function becomes the only legitimate description of any system.

However, who is doing the mapping to make a particle's position real? We can imagine a particle such as an atom in an empty infinite space where it has no definable position. It can be said to exist as a superposition of all possible positions within that space. An external object is placed alongside the atom. Now the atom's position can be mapped with reference to the object. Will it remain as a superposition of all possible positions? If the external object is a scientist equipped with an electron microscope, then she can measure the atom's position and thereby force it to adopt a real position. What if the external object is a cat, or a microbe, or a rock? Can these objects measure the atom's position? What if the cat is blind? We can see that this is bringing us back to the lair of Schrodinger's cat. However, this way of looking at the problem does reveal an essential feature of quantum measurement: there must be *some exchange of matter or energy between the quantum system and its environment/measuring device*. If a rock is merely next to the atom, how could the rock and atom *know* of each other's existence? For the rock to know the atom there must be some sort of exchange of matter or energy between them. For the rock to *measure* the atom's position, some of the atom's possible positions must make a difference to the rock. Perhaps in one particular position the atom will release a photon on a trajectory that will impact with the rock, which happens to be poised on a ledge, so that the tiny momentum delivered by the photon is sufficient to topple the rock. They key point is that for a quantum measurement to take place there must surely be some kind of record of the quantum system (the atom) held in the environment (the rock). There must be some leakage of information about the system into its environment.

But what if the object placed alongside our atom is another fundamental particle such as an electron. As discussed above, the new particle can become entangled with all the possible positions of the original particle as an EPR-pair. No measurement can take place. A rock may be capable of measuring a particle's position but a lone electron would not. What about two or three electrons, or a million, billion or a trillion electrons inside a trillion atoms? Clearly we will eventually be reaching the kind of number of atoms within the average person – or cat (somewhere in the region of 10^{30} atoms). Between an atom, rock, cat and person there must be some point of transition where quantum entanglement gives way to classically separated states. Where is this point?

To answer this question we will return (for the last time) to the two-slit experiment. When light was shone through the slit screen we saw an interference pattern. However, it is actually quite tricky to obtain an interference pattern. In our experiment we used a laser as a light source because it releases *monochromatic* light (light of a single frequency or colour) *in phase*. The critical feature of monochromatic laser light is that the waves all have the same wavelength and are emitted in the same phase: the peaks and troughs of the waves all march along in step. Only this kind of *coherent* light will generate the sharp strips of light and dark bands that characterize the interference pattern. If instead polychromatic light is used, which is made up of a mixture of light of different wavelengths, the waves quickly fall out of step, and become *incoherent*. When the waves arrive at the screen, the peaks and troughs are mixed up, arriving at different times. The more or less random addition and subtraction of the incoherent peaks and troughs leads to the cancellation of all the quantum superposition $\{RIGHT\ (+/-)\ LEFT\}$ terms. Only the simple $[RIGHT] + [LEFT]$ scatter pattern remains, with no interference bands.

Another way to disturb light's coherence is to jiggle the photons around a bit. The argument will be clearer if at this stage we switch to firing electrons through the slit to generate an electron interference pattern. As with photons, the interference effects indicate that each electron is able to travel through both slits as a quantum superposition. The electron *lives* in the quantum world. We next fire photons at the electrons to jiggle them about. Some electrons will gain momentum from the impacts; others will lose it. The net result is a loss of coherence for the electron beam (*decoherence*) and destruction of the interference pattern. But isn't this precisely how we detect electrons? The kind of detector we use to catch electrons in their flight, fires photons at the electrons, measuring their deflection. But we have so far accepted the loss of interference as evidence of quantum measurement: collapse of the wave function. Yet it turns out that – at least in this situation – we do not need the collapse hypothesis to account for the phenomenon. Applying standard quantum mechanics principles to both the electrons and photons demonstrates that decoherence alone leads irrevocably to the loss of interference. With the loss of interference effects, we have also lost the evidence for electrons going through both slits. For all practical purposes, the electrons now live in the classical world. It is as if a quantum measurement has taken place, but without the intervention of any Copenhagen-style observer. The environment (the photons) has effectively measured the quantum system.

It is important to note that the evidence for electron interference has not entirely vanished. But where has it gone? It has been carried off by the photons. If all of the photons that impacted the electrons were trapped and analysed, the evidence for interference would still be discernible. But it is virtually impossible to catch all the photons and analyse them. The evidence for electrons passing through both slits may still exist within the world, but for all practical purposes (or FAPP, as John Bell puts it), a quantum measurement has taken place and the quantum world is lost to the experimenter.

Wojciech Zurek working at California's Institute of Technology has proposed that the reason why the entire world appears classical to us is because of decoherence. Just as an electron beam may be bombarded with photons, so any open system is continually bombarded with photons, electrons, atoms and other particles. The quantum system will inevitably become entangled with the fate of billions of particles in its environment, and this entanglement will cause decoherence. Interference effects will be erased, so that the world will appear classical. The environment will *measure* the quantum system.

In decoherence models, the environment need not be limited to the external environment but can include the internal environment of the quantum system – its 'degrees of freedom'.[22] Metal rods held at temperatures close to absolute zero, known as Weber bars, are used to detect gravity waves. The rigidity of Weber bars, combined with freezing temperatures, induces the atoms to vibrate in unison – coherently. In this state, the Weber bar can display quantum interference effects. However, if a bar is heated, the thermal motion of its atoms and molecules, jiggling and bumping about, randomizes the phase of the atomic vibrations so that they no longer vibrate coherently: decoherence. Interference effects are no longer detectable. If the bar were made of more flexible material, then the random motion of atoms and molecules within the bar, even at low temperatures, would similarly lead to decoherence. Decoherence is the same as with the electron beam but it is now the internal environment of the quantum system – rather than its interactions with an external environment – that is doing the jiggling. As with the electron beam, the interference effects have not vanished from the world. If any particle within the bar were examined, then it would continue to attest to its ability to exist as a quantum superposition. It is only when we examine the entire bar that the interference effects cancel each other out. For all practical purposes, the interference effects have vanished and the bar behaves classically. However, once again it should be remembered that it is only the *evidence* for *quantumness* – the interference effects – that have vanished from the world. Individual particles still exist, as quantum superpositions; we just don't see them. The quantum weirdness is hidden, but is still there.

The key to decoherence is the involvement of interaction between the quantum system and a complex environment. After the measurement, the environment must hold some kind of indelible record of the quantum event. The record could be carried off into space on the backs of photons, or could be dissipated amongst a billion particles within the quantum system itself. But for decoherence to cause quantum measurement, there must be some *leakage* of information into the environment.

Decoherence provides an explanation for why we never see the evidence for quantum superposition in big objects. An object as big as

Schrödinger's cat would rapidly succumb to decoherence. Its atoms and molecules have billions of degrees of freedom, interacting with each other and with the external environment. The quantum superposition of both live and dead cat would be swiftly abolished by decoherence. Decoherence also explains why the thermodynamic processes examined in Chapter Six never display quantum effects. Steam engines and chemical engines driven by *incoherent motion* have billions of degrees of freedom. The bumping and jiggling of their atoms and molecules destroys interference effects and so causes decoherence. Most natural phenomena, such as the weather or the motions of the planets, are also driven by incoherent motion and fall prey similarly to decoherence. Even self-organizing structures such as convection flow, anticyclones or the Belousov-Zhabotinsky chemical reaction are, at the molecular level, driven by incoherent motion. Quantum effects are washed away by decoherence.

So does decoherence solve the measurement problem? Not entirely. For one thing, it does not strictly lead to the same result as quantum measurement, *à la* Copenhagen. We earlier abbreviated quantum measurement in the two-slit experiment to: ψ[{photon at RIGHT} (+/−) {photon at LEFT}] + *Qmeasurement* → ψ[photon at RIGHT detected by right detector] *or* ψ[photon at LEFT not detected by right detector]. This correctly describes the situation that, before a measurement, the system was in a quantum superposition, but after measurement, wave-function collapse forced the system to choose a single state. Decoherence generates an almost identical situation: ψ[{photon at RIGHT} (+/−) {photon at LEFT}] + *Decoherence* → ψ[photon at RIGHT detected by right detector] *and* ψ[photon at LEFT not detected by right detector]. Just as with the wave-function collapse equation, the quantum-mechanical, (+/−) term that gave rise to the interference effects has disappeared after decoherence.

The departure of the (+/−) interference term allows us to ignore the influence of quantum superposition, just as if a quantum measurement had taken place. The only difference between this and the standard wave function collapse is that decoherence leaves us with the two possible outcomes linked by *and* rather than *or*. When many electrons are fired through the screen in the two-slit experiment, then we obtain what is termed a statistical mixture of states with some of the electrons predicted to travel through the left slit *and* some through the right. By allowing the environment to measure the quantum state, we have a perfect agreement between theory and experiment, but now entirely within the framework of Schrödinger's equation, without having to resort to any arbitrary *wavefunction collapse* hypothesis.

The problem comes when we try to use decoherence to predict what happens to a single electron. The decoherence-reduced measurement equation gives us precisely the same answer as for an ensemble of electrons: ψ[photon at R detected by R detector] *and* ψ[photon at L not detected by R detector]. The interference terms are gone, which is fine. It is the remaining *and* that is the problem. How can a single electron go through both the left *and* the right slit? This brings us back to the same thorny problem of quantum superposition. If we try to measure the electron then we will find it inhabits only a single state. But the decoherence equation describes a measurement that includes both. What has happened to the state that has been discarded?

Many physicists take the stance that quantum mechanics is solely about statistics; it has nothing to say about the fate of individual particles. Since these physicists often also claim that quantum mechanics encapsulates everything that can be said about the world, they often go on to claim that questions about individual particles are therefore meaningless. However, this is becoming increasingly untenable as experiments are beginning to probe the realm of single particles. Researchers in the rapidly advancing field of nanotechnology have already developed techniques for manipulating individual atoms to construct ultra-microcircuit boards. And of course (as we discovered in earlier chapters) living cells learned to manipulate single particles billions of years ago. The cells of those physicists who deny the reality of events involving single particles are living proof of the absurdity of their claim.

Physicists who do worry about the problem, or have to deal with it, tend to look to the Many Worlds interpretation. When a single electron passes through two slits, the universe splits. If no attempt is made to discover the electron's route, the two universes somehow inhabit the same space and the two particles interfere with one other. However, if you attempt to detect the electron path, your measurement causes decoherence, the universes separate, and the electrons go their own way in separate universes. The universes no longer interact, eliminating interference. You can discover which universe you happen to inhabit by observing whether your electron went through the left or right slit.

It is generally assumed that the universe splitting coincides with decoherence, but this does not necessarily follow from the decoherence equations. Remember that decoherence destroys the *evidence* for quantum superpositions, the interference effects, not necessarily the superpositions themselves.[23]. Superposition may persist beyond decoherence, but leave no trace (for all practical purposes). Finally, it should be remembered that decoherence is not exclusive to the Many Worlds interpretation. One could just as easily invoke a Copenhagen-style wave function collapse to eliminate the excess states left by decoherence.

Although the phenomenon of decoherence is certainly real and is one reason why the world appears *normal* to us, it cannot be the only one. It does, however, give us a tool we can use to find the border between the quantum and classical worlds. The next question to address is, of course, at what level within living cells does that border lie?

10

The Beginning

The natural temptation is to dismiss the strange and unfamiliar elements of quantum mechanics as illusion and claim, 'It's all nonsense; the world is just not like that!' But quantum mechanics has stood the test of time and the rigours of thousands of experiments for nearly three quarters of a century. It underpins the technology of our modern world in applications as varied as computers, telecommunications, lasers, compact disks and medical imaging. If quantum mechanics were fundamentally wrong, then it is likely that we would know its flaws by now. Our computers would not work; our CDs would not play; and experiments would give inexplicable results.

Perhaps, someday, an experiment will be performed that contradicts quantum mechanics, launching physics into a new era, but it is highly unlikely that such an event would restore our classical version of reality. Remember that nobody, not even Einstein, could come up with a version of reality less strange than quantum mechanics, yet one which still explained all the existing data. If quantum mechanics is ever superseded, then it seems likely we would discover the world to be even stranger.

In the absence of contradictory data or any viable alternative hypothesis we must, perhaps reluctantly, accept that, at the atomic level, the constituents of matter dance to a different tune from that played on a classical violin. The protons, electrons, atoms and molecules that form every printed dot on the page you are reading must obey quantum not classical laws. The atoms that make up your eye, your brain, your hands and the matter of every other living organism, must similarly obey quantum laws. The reason the world appears *normal* to us is that those macroscopic inanimate objects big enough for us to see are made of billions of atoms and molecules, all bumping and jostling about. This muddle of incoherent motion dissipates the quantum phenomena of coherence, superposition and uncertainty. The quantum level of reality is still there, but hidden behind a veil of decoherence.

However, there is one macroscopic phenomenon not driven by the incoherent motion of multitudes of particles: life. We discovered in Chapter Three that the motion of electrons and protons within DNA initiate the mutations that drive evolution. Indeed, the structure of a single molecule – DNA – drives the actions of every cell inside our bodies and other living organisms. Nothing else in the world we see around us (or in the heavens above) is similarly contingent upon dynamics at this level. We examined in Chapter Five how the enzymes inside living cells perform their actions by mobilizing single electrons, protons, atoms and molecules. Single protons are batted across membranes to power the molecular turbine engines of mitochondrial respiration. Molecular proton guns initiate muscle contraction. And fundamental particles do not obey classical rules. Life must, instead, dance to the quantum fiddle. Let us next explore the consequence of this realization for our understanding of the phenomenon of life.

QUANTUM MEASUREMENT AND THE ORIGIN OF LIFE

Biology's biggest mystery is its first: the origin of life. In Chapter Four, we looked at how the development of modern-day plants, animals and microbes from the first living cell is broadly understandable in terms of neoDarwinian evolutionary theory. Similarly, the earlier emergence of a living cell from the first self-replicating entity does not pose any major challenge to the standard dogma (although the rapidity of this step is still a puzzle). The biggest problem is to delve deeper in time to account for how the first self-replicator arose. Darwinian natural selection works only above the level of things that can already self-replicate. We cannot use it to explain the spontaneous emergence of self-replicating organisms. Yet, as discussed in Chapter Four, standard thermodynamic chemistry seems woefully inadequate to fill the gap between the primordial soup and the minimum complexity required for self-replication.

What is lacking in test-tube thermodynamic chemistry is directionality. The incoherent motion of billions of particles inevitably leads any complex mixture of chemicals into a maze of chemical pathways and a jumble of products – the inevitable *gunk* of primordial soup experiments. Totally unfeasible tropical forest-sized yields of complex organic products are required to make even a single self-replicating molecule. Yet, as we discovered in Chapter Five, the product of life's emergence, the living cell, is driven by a different principle from standard thermodynamic chemistry – the directed motion of individual particles. A chemistry of small numbers, a *quantum chemistry*, is needed to account for life's emergence.

To see how quantum mechanics might have been involved in life's origin, imagine once again the primordial soup described in Chapter Four: perhaps the 'warm little pond' envisaged by Darwin. At this stage we will not worry about how that soup was generated or where it was situated. We will assume that some standard chemistry – perhaps Miller-Urey-type chemical reactions (Chapter Four) taking place in a sulfurous pool or within a deep ocean spring – made the soup. The soup's ingredients would have included many simple organic compounds such as amino acids, simple sugars and perhaps even nucleotides. Let us also suppose that there is some way of putting these components together to generate a self-replicating entity. Since amino acids are at least plausible prebiotic chemicals, we will assume, for the sake of simplicity, that the first replicator was a short peptide. Perhaps it was like the 32 amino acid *self-replicating* peptide engineered by Lee, described in Chapter Four (although it must be admitted that the conditions necessary for replication of Lee's peptide are very far from being plausibly prebiotic). The self-replicating peptide can be written in one letter amino acid code[1] as: *RMKQLEEKVYELLSKVACLEYEVARLKKLVGE*. As a further sacrifice to simplicity let us assume that conditions in our warm pond promoted peptide bond formation – one amino acid sticking to another – but only allowing the peptide bonding found in natural proteins. These assumptions may seem somewhat of a cheat and we will return to re-examine them later. They do, however, allow us to concentrate on the crux of the problem for the origin of life: how was the first self-replicator assembled? There are 20^{32} or 10^{41} possible ways to put together peptides which are thirty-two amino acids long. If primordial chemistry managed to synthesize just one molecule of each of the possible peptides, then the total mass of peptides must have weighed about 10^{18} kilograms, far more than the total mass of organic carbon in today's tropical forests. There's no warm pond big enough to make so many peptides.

A graphic way to visualize amino acid additions is as a walk through a twenty-dimensional space,[2] with each dimension representing one of twenty possible amino acids. We start with the first amino acid (arginine for our model self-replicator) as a point in space, and each successive amino acid addition represents a single step in one of the twenty possible directions. To synthesize a particular peptide we need to trace a single path through *amino acid sequence space*. To make the thirty-two amino-acid self-replicating peptide, our walk is a short trot of thirty-two single steps. It sounds easy. The problem is that a twenty-dimensional space represents an enormous volume of sequence space. Consider how the quantity of space increases in going from a one-dimensional line to a two-dimensional plane and then to a three-dimensional box. Each extra dimension brings a huge increase in available space. Now consider adding seventeen extra dimensions to a box and you get some idea of the vast amount of sequence space a growing peptide has to navigate. A single thirty-two-amino-acid peptide represents only one very fine track – one of 20^{31} equally probable thirty-two-amino-acid-long tracks that lead off from arginine. The problem for biogenesis is to *find* the right path to the self-replicator.

Before starting on our amino acid trek, it is useful to recall how Darwinian natural selection solved similar problems for living organisms. The evolution of the eye, briefly described in Chapter Four, can also be represented as a walk in multidimensional space, only now the number of dimensions (representing all possible kinds of modifications to eye structure) are vast. The light-sensitive patch on the surface of a single-celled protozoan can be represented as the starting block, and the path to the functional mammalian eye as only one fine track in a gigantic volume of eye-design multidimensional space. Evolutionary advance along the track is driven by mutations, but they are generally held to be random (this claim will be re-examined in the next chapter), and are therefore represented as hops in every possible direction. Evolving with mutations alone is akin to relying entirely on thermodynamic chemistry for primordial synthesis. Both processes lack directionality. But natural selection rescues Darwinian evolution. Only those offspring fit enough to thrive, prosper and generate more offspring, continue the trek. Most of the fit will be very similar or identical to their parents but very occasionally some mutants will arise that have acquired a more effective eye; and their ability to see that little bit better will drag their offspring in the better eye direction. In this way, the combination of a random process (*mutagenesis*) and a directional process (natural selection) steers evolution along the path towards increased fitness.

The crucial prerequisite for Darwinian evolution is that natural selection, acting at each and every step along the path, serves to *capture* mutants with increased fitness and so provides the direction of the evolutionary walk. Now returning to our peptide walk, we can see that what we need is some mechanism that can similarly guide the evolution of the growing peptide along the path that leads from arginine to the peptide *RMKQLEEKVYELLSKVACLEYEVARLKKLVGE* when there are 20^{31} minus 1, possible wrong turnings. To solve this puzzle we must once more abandon classical logic and *think quantum*.

Let us start with our first amino acid at point R (for arginine) in the multi-dimensional peptide sequence space. We will imagine that, in our primordial soup, a single molecule of arginine had just enough energy to react with another amino acid that might have been another arginine or one of nineteen other amino acids, to make a *dipeptide*, RX (X referring to any of the twenty amino acids). Classically, the arginine molecule must make a choice:[3] which amino acid do I react with? In sequence space, this choice amounted to a single step in one of twenty possible directions. However, quantum-mechanically, no choice need have been made. The single arginine molecule would have reacted with all twenty possible amino acids as a quantum superposition. The resulting dipeptide would then have existed as a quantum superposition of all twenty possible RX dipeptides. In multi-dimensional sequence space, our peptide addition took a step in all possible directions as a quantum superposition.

The quantum superposition need not have stopped at the dipeptide stage. Another amino acid might have been added to make a tripeptide which again existed as a superposition of all four hundred (20^2) possible R-X-X tripeptides. The tripeptide superposition may have similarly reacted to generate a superposition of all eight thousand (20^3) possible

R-X-X-X tetrapeptides. The process could have continued to add more and more amino acids until a peptide which was a superposition of all 10^{41} possible thirty-two-amino-acid peptides that start with arginine was eventually generated, *so long as the system remained at the quantum level.*

QUANTUM COHERENCE IN THE PROTO-CELL

But can a peptide be in a quantum superposition? The answer will depend, to a certain extent, on the interpretation of quantum mechanics you prefer. Many physicists who choose the standard Copenhagen interpretation consider that the entire universe must exist as a quantum superposition of all possible states, until a conscious observer makes a measurement. Anyone prepared to consider that even a cat could exist as a quantum superposition of being both dead and alive, should have no problem with a superposition of a thirty-two-amino-acid peptide. Indeed it is difficult to see how any quantum measurement *à la* Copenhagen, could have taken place on the early Earth – perhaps our alien spacecraft could have visited and done the trick?

However, we discovered in the last chapter that decoherence provides us with an alternative means of delimiting the effects of quantum mechanics in physical systems, without reference to any external observer. When a quantum system couples with its environment, decoherence destroys interference effects and – at least the evidence for – quantum superposition. The environment *measures* the quantum state. The question then arises: when did decoherence intervene in the primordial soup? You will recall that decoherence depends on information about a quantum system leaking into its environment. Just how much information was needed to leak into a peptide's environment to destroy quantum coherence is difficult to answer. It is likely to have depended on how *isolated* the peptide was from its environment.

Warm, wet ponds do not appear initially particularly promising systems to isolate quantum superpositions. However, one way to preserve quantum coherence is to keep the quantum system small and isolated. Cees Dekker, and colleagues at Delft University of Technology, have demonstrated that electrons within ultrafine carbon capillaries (carbon nanotubes) remain quantum coherent throughout the tube's entire length. Admittedly, this was achieved at temperatures close to absolute zero ($^-273°C$). However when another quantum level phenomenon, *superconductivity*, was first discovered, it was also thought to be restricted to temperatures close to absolute zero. Superconductivity is quite different from the electrical phenomenon which powers our televisions and washing machines. Standard electricity is a classical phenomenon; a manifestation of the motion of billions of electrons that move through wires in much the same way as gas particles move through space to fill an empty box. The motion of individual electrons is entirely incoherent; it is only the bulk of electrons that exhibit an average motion in the direction of the current. However, in 1911, the Dutch physicist Kamerlingh Onnes discovered that, at temperatures close to absolute zero, the conducting electrons within certain materials lose their individual identity and travel as a single coherent pulse of electrical charge. Even more remarkable was the discovery in 1986 that some mercury-based compounds became superconducting at temperatures as high as $^-89°C$. Glimpses of superconductivity at even higher temperatures have also been reported.

High-temperature quantum phenomena are not restricted to exotic materials. The unusual reactivity of benzene, one of the world's most widely used chemicals, depends upon the ability of its *pi* electrons to *delocalize* amongst the six carbon atoms of its molecular ring structure. Delocalization is just a chemist's way of saying that the electrons are in a quantum superposition of position states around the benzene molecule. Even more remarkable was the discovery (already mentioned in Chapter Seven) of quantum interference effects for fullerene which showed that individual C_{60} molecules can pass through both slits of the two-slit experiment as a quantum superposition. But a simple metal wire can also demonstrate quantum effects. If the wire's electrical resistance is monitored whilst being slowly broken, before it finally snaps, its resistance takes on quantized values in units of Planck's Constant. It seems that, at the final moments before the wire is torn apart, the electrical conductivity proceeds only through a very small number of atoms and thereby obeys quantum laws, even at room temperature.

So quantum level effects may be found at ambient temperatures or higher (the fullerene interference experiment was performed at a scorching 600°C), particularly at very small scales. There were probably many opportunities for primordial chemistry to become trapped inside tiny structures: perhaps inside the pores of a rock or within a chemically generated oil or protein droplet. These nanoscale structures would have served as a kind of proto-cell (or *very* small warm pond) which might have protected the coherence of the quantum states inside. A peptide forming inside such a proto-cell would have been shielded from the coupling with a complex environment that would normally lead to decoherence. Its quantum state might well have remained intact – at least transiently – superpositions and all. New molecules, including a fresh supply of amino acids, would have diffused into and out of the proto-cell and reacted with the trapped peptide to add to the growing chain. If the growing peptide remained trapped within the proto-cell, then the product of the reactions would be invisible to the proto-cell's surroundings. No information about the exact chemical additions would have leaked into the environment, so decoherence would have been avoided. Instead of a classical addition of a single amino acid to make a single peptide product, peptide addition would have taken place as a quantum superposition of all possible peptides.

QUANTUM MEASUREMENT IN THE PROTO-CELL

But peptides are not life. We still need to evolve a self-replicator. This is where quantum measurement comes in. The proto-cell would not have been able to shield the quantum system from its environment indefinitely. There must have been a point when the environment *noticed* what was going on and *measured* the quantum system. When did this happen? The conscious observer Copenhagen*ist* viewpoint is that three and a half billion years of history would have passed before that measurement was performed. Decoherence gives us a more acceptable observer-free criterion for quantum measurement. Decoherence would have occurred when *the quantum system coupled irreversibly with a complex environment*. The molecules that most effectively couple with their environment are *enzymes*. Enzymes transform thousands of molecules of substrates into products, and so change the positions and energies of millions of particles in their environment, causing decoherence.

But can a peptide be an enzyme? We discovered in Chapter Five that enzymes are (normally) very large proteins made of strings of hundreds of amino acids. However, although proteins are the most powerful enzymes, short peptides also have enzymatic activity. A powerful new approach to *designing* new enzymes, called *combinatorial chemistry*, is to screen *libraries* of billions of short random peptides for enzymatic activity. It is not usually difficult to find a peptide with a particular enzymatic activity (albeit weak compared with protein enzymes) in a library of a million random peptides, only six amino acids long.

So our quantum superposition of a peptide trapped inside the proto-cell would inevitably have contained many quantum proto-enzymes with weak enzymatic activities, capable of generating classical signals. These proto-enzymes would have coupled with the environment and so collapsed the quantum state of the peptide into a classical state. However, and most importantly, *whilst the peptide remained a single molecule, it could always re-enter the quantum realm after the measurement.* Although the proto-enzyme's interactions with its environment might have modified the position and energies of many particles, it would have emerged unscathed from the measurement process (recall from Chapter Five that enzymes are not irreversibly modified in enzymatic reactions). Thereafter, it would have been free to drift once more into the realm of quantum superposition and await the next measurement.

This process of drifting into the quantum realm, measurement, collapse into a classical state, and drifting back into the quantum realm, would have continued as the peptide added more and more amino acids. Some amino acid additions – those that did not confer enzymatic activity – would have taken place in the quantum realm, whereas others would have precipitated quantum measurement and a brief return to the world of classical physics. This process would have continued to elongate the quantum superposition of possible peptides until such a time when the system irreversibly collapsed into a classical state. The point at which this irreversible collapse would have taken place is easy to predict: it would have been when the peptide learned to self-replicate.

Once the growing quantum peptide chain hit upon a proto-enzyme that managed to replicate itself, then it would have inevitably and irreversibly amplified its state to the classical realm. In contrast to other kinds of enzymatic activity that leave the enzyme unscathed, self-replication cannot leave the enzyme unscathed – it is replicated. Each replicated copy of the proto-enzyme would have identical enzymatic activity to its parent and would make more self-replicators. Each cycle of enzymatic replication would have changed the position and energy of many particles within the proto-cell and an army of self-replicators would change the positions and energies of billions of particles. While a single peptide enzyme molecule could remain in the quantum realm, a self-replicator inevitably amplifies itself to a classical entity. Some self-replicators would have leached out of the proto-cell, initiating similar changes in the proto-cell surroundings. Information *about the peptide* would leak into the environment to cause decoherence and irreversible quantum measurement. Quantum superposition would have been shattered and the self-replicator would have crashed out of the quantum superposition of billions of possible peptides. Thus, the emergence of the self-replicator *nailed* the growing peptide chain to a classical reality.

THE ANTHROPIC MULTIVERSE ORIGIN OF LIFE

In essence, the self-replicating peptide within the quantum superposition of peptides would have performed a quantum measurement on itself. But there is a catch (life is never that easy). Out of the quantum superposition, a single classical peptide would have emerged. *But, there is only a very small probability that the single peptide that crashed out of the superposition would have been the self-replicator.* Remember that null measurements are perfectly good quantum measurements. A detector placed over one slit in the two-slit experiment is sufficient to collapse the wave function, even though it may not detect a photon. The mere possibility of its detecting a photon is all that is needed for quantum measurement. In the same way, the mere possibility of a self-replicating peptide within the superposition of possible peptides would have been sufficient to perform a quantum measurement. Though the presence of the self-replicating peptide within the superposition was the critical factor that destroyed the quantum state, the chance that this self-replicator emerged from the superposition would have been no more than that of any other peptide: one in 20^{32}.

So we are no farther forward. We are left stranded with only one random thirty-two-amino-acid peptide that is (very probably) useless at self-replication. All the trillions of other peptides that formed the superposition – including the self-replicator that caused its demise – have vanished. But where have they gone? The usual recourse of adherents to decoherence is to invoke the quantum multiverse. In the Many Worlds interpretation, there would have been a separate universe created for every one of the 20^{32} possible peptides that emerged from the superposition of peptide states. The self-replicating peptide would materialize in only one of those 20^{32} universes.

I must admit that my first response to the Many Worlds hypothesis was total incredulity. The idea that the universe splits into billions of parallel universes whenever a quantum event is forced to make a choice is preposterous and seems an incredible waste of universes. Now, I am not so sure. The great advantage of the hypothesis is that, once you accept it, the interpretation of quantum phenomena is much simpler. It is far easier to work out what is going on in quantum mechanics with a multiverse to partition the outcomes, rather than to rely upon the more nebulous concepts of *wave-function collapse*, hidden variables, pilot waves, or backward-in-time signalling. And there are many very eminent physicists (particularly amongst cosmologists), including Nobel laureates, who believe adamantly in the multiverse. So, to exploit the simplicity of Many Worlds, I will stick with it henceforth in the following pages, only occasionally expressing a concept in different terms. If you still find the multiverse hard to swallow, don't worry. You can consider it as merely a convenient metaphor for some deeper form of reality. Remember that all of the *conventional* interpretations of quantum mechanics generate precisely the same predictions and therefore with existing data are indistinguishable. Any conclusions I reach with the multiverse can be translated into whatever interpretation you happen to prefer.

So, using the Many Worlds approach, peptide addition within the proto-cell would have taken place, not in a classical universe, but within the multiverse of all possible states. Each peptide addition was a multiple branch point where the growing peptide chain evolved in all directions simultaneously. The multiverse expanded with every addition to the peptide chain as the quantum tree of possibilities grew simultaneously in every direction. But once one branch lighted upon a self-replicating peptide, then quantum measurement would have become inevitable. Decoherence shattered the unity of the quantum tree, which split into 20^{32} separate branches, each representing a different universe. Only one of the 20^{32} descendent universes harboured the self-replicator, but in that universe life emerged.

But, in which of those 20^{32} separated universes that collapsed out of the peptide superposition do we live? It could only be the one in which the self-replicator emerged. One secure fact we know about the universe we inhabit is that it generated self-replicating entities. We *must* inhabit the one lucky universe amongst the other 20^{32} minus one universes, because we are here to talk about it. This Anthropic Multiverse hypothesis can be considered as a variant of the anthropic principle, discussed in Chapter Four. Just as our existence guarantees that the laws of physics and the fundamental constants are compatible with biology, so it also guarantees that our universe was lucky enough to draw the winning peptide out of the multiverse.

The *anthropic multiverse* principle neatly sidesteps the problem of how prebiotic chemistry took place within a small warm pond, yet still manages to explore the vast array of possible peptides (or RNA molecules, or any other prebiotic structures) to find the first self-replicator. In this view, prebiotic chemistry took place inside a proto-cell inhabiting the entire multiverse – there 20^{32} was still a tiny number. The Anthropic Multiverse also circumvents the additional chemical problems skipped over in our simplified peptide model of prebiotic evolution. If you remember, we allowed only certain chemical reactions to take place in the primordial soup – those that led to peptide bonds. The only difference this extra complication makes to our multi-dimensional walk within the multiverse is to place each peptide addition within the much larger number of dimensions that (must now) encompass every possible chemical reaction. Allowing greater chemical flexibility will increase the number of possible structures in the superposition, but will only expand the number of possible universes that must be accommodated. Substituting RNA, or any other prebiotic biomolecule, for the first replicator is also perfectly compatible with this quantum evolution within the multiverse. The evolving sequence space would have been a space of polynucleotides, rather than peptides, but quantum measurement would have similarly caused RNA self-replicators to emerge from the quantum multiverse.

If you are still unhappy with the multiverse, you can *reinterpret* the quantum evolution of peptides in the language of your choice. The Hidden Variables interpretation of quantum mechanics would have the pilot wave travelling through all possible peptide structures to signal the presence of the self-replicating peptide which could perform quantum measurement. In the transactional interpretation, quantum measurement (by a self-replicator) would have sent a signal that travelled backwards in time to the growing peptide. The Copenhagen interpretation would either have to accept that living cells could act as quantum-measuring devices, or would have had to await the arrival of conscious observers to provide the measurement. In this guise, the anthropic multiverse hypothesis becomes very similar to the theory proposed by the physicist John Wheeler of a *participatory universe*, whereby our presence as observers (who perform quantum measurements) *made* the universe we inhabit *real*. Each interpretation represents a different way of thinking about the course of events, but all point in the same direction. In each case, the earliest steps in life's emergence took place at the quantum level where primordial chemistry could explore a vast array of multiple states simultaneously. Quantum measurement by a self-replicating biomolecule was the key event that caused life to emerge.

The anthropic multiverse provides an explanation of how we came to be here, but I must admit to as much dissatisfaction with it as with the original anthropic principle. For one thing, it only works once. To see why this is so, imagine another warm pond, perhaps beneath the ice of Europa. Millions of years ago, the same kind of amino acid quantum chemistry may have taken place under the ice, to generate a superposition of all possible peptides. A self-replicating peptide would have emerged within the superposition, to perform quantum measurement and precipitate itself out of the multiverse. Darwinian natural selection would have taken over; and, billions of years later, alien fish would be swimming beneath Europa's ice. *But not in our universe.* The Europaean peptide's quantum measurement would have caused life to emerge in one of the 20^{32} different universes which fell out of its quantum superposition, but there is only one chance in 20^{32} minus one that our universe would strike lucky on this second occasion. We can no longer use our own presence as an anthropic selection, because we're already here. The *Europaean* self-replicator and its descendants would be somewhere in the multiverse, but there would be only one chance in 20^{32} that we inhabit the same universe. The consequence of this reasoning is that while life may be sprinkled throughout the multiverse, within any single universe it is likely to be unique. So although the anthropic multiverse can account for the presence of life on Earth, it could never account for an independent origin of life anywhere else.

If, like me, you would dearly like to believe that we are not alone in the universe, then the anthropic multiverse hypothesis makes for depressing reading. Yet, it must be admitted that, apart from the (not particularly convincing) Martian meteorite microfossils, there is no evidence for life beyond our planet. Sadly, the anthropic multiverse theory is consistent with the existing data. The principle would also make depressing reading for any prebiotic chemist who hoped to win his Nobel prize by demonstrating the spontaneous emergence of life in the laboratory. Although able to simulate the prebiotic quantum chemistry that gave rise to life, he would nevertheless not see life emerge. His laboratory-grown primordial soups might contain quantum superpositions of peptides but, once again, the self-replicating peptide would be found in only one of 20^{32} universes that pop out of the superposition. In all the other universes, only more gunk would be generated. It is highly probable that our stalwart chemist would find himself inhabiting one of the gunk universes (though obscurely comforted by the thought that one of his 20^{32} multiverse twins would be shouting 'Eureka!'). Once again, we must reluctantly accept that an anthropic quantum origin for life fits the evidence.

However, despite its inevitability and its consistency with the facts, I still find an anthropic quantum origin, deeply unsatisfying. This is, admittedly, mostly for selfish reasons. I very much hope that one day I will see pictures of real extraterrestrial lifeforms on my television. I hope one day to witness self-replication emerge in a laboratory somewhere. But is there an alternative explanation – some way to load the quantum dice so that self-replication is more likely to emerge out of the multiverse? I believe there is.

ZENO AND THE ORIGIN OF LIFE

Think about the multidimensional walk that led to the self-replicator. It should be familiar in that it strongly resembles the way the inverse quantum Zeno effect can direct a quantum system along a particular path. If you remember, the inverse quantum Zeno effect represents one of the peculiarities of quantum measurement, whereby a dense series of measurements along a particular path can draw that system along that path. Insertion of extra polaroid lenses, between vertically and horizontally polarized lenses (a perpendicular or orthogonal pair) rotates the angle of polarization of light. Without the extra lenses, the distance between the two states (vertical and horizontally polarized light) is too great. No light gets through. But the insertion of the extra lenses – each performing a quantum measurement – lays down a series of stepping stones which rotates the light from one state (vertically polarized) to another (horizontally polarized) in a series of short *hops*. A sufficiently dense series of measurements will lay down a path that will allow the photons to evolve smoothly from one state into another. Similarly, a dense series of quantum measurements of a particle along a positional path can move the particle along that path. The path may be only one out of a trillion equally probable, but quantum measurement can force the system to evolve in that measured direction.

But isn't this more or less what we want to do when we evolve the single amino acid arginine (R), along a single path through peptide multi-dimensional space to synthesize the self-replicator, *RMKQLEEK-VYELLSKVACLEYEVARLKKLVGE*? In classical terms, the probability of the peptide taking the right route is close to zero (in fact $\frac{1}{10^{32}}$). But if quantum measurements were performed along the route, then the inverse quantum Zeno effect could make that path much more likely. Can the inverse quantum Zeno effect make the reaction more probable? To see how, recall how the inverse quantum Zeno effect actually works. It depends on the ability of oblique quantum measurements to *decompose* a quantum state into sets of orthogonal (perpendicular) states. Measurement then forces one of those states to become real, and a dense series of measurement forces the system along the measured path. The polarization state of photons can be rotated by the inverse Zeno effect, as can the position of particles. If positional measurements can be performed to move electrons around in empty space, they can equally be performed to move electrons (or protons) around in the space of atoms and molecules.

Looked at this way, the chemical reaction leading to the first self-replicator (or indeed any chemical reaction) becomes a sequence of electron and proton *movements* within and between molecules. The walk in multi-dimensional peptide addition space, that led to the self-replicating peptide, becomes a walk in position space for electrons and protons. The route to a single self-replicator will still be extremely improbable but quantum measurement and the inverse Zeno effect could *capture* the motion of the particles along that route and thereby make it more likely.

But how can the motion of particles, within and between molecules, be measured? As discussed above, enzymes can perform that feat. The precise position and energy of particles within enzymes is crucial for their enzymatic activity. Enzymes regularly perform quantum measurements of the states of their own particles. I have already described how proto-enzymes would have inevitably emerged in the quantum peptide. The enzymatic actions performed by these proto-enzymes would have performed quantum measurement of the particles that make up those peptides. The proto-enzymes would have emerged unscathed from the measurement process and – after measurement – their quantum state would have drifted back into the quantum realm. However, any chain of electron and proton motion that led to a self-replicator would have been irreversibly amplified into the classical world. Although proto-enzymes may remain at the quantum level, a self-replicator inevitably amplifies its quantum state to the classical level. The self-replicator nailed the growing peptide to the classical world.

We are then left with a chain of quantum measurements terminating with a self-replicator that amplified the quantum system to the classical level. This is exactly analogous to the inverse quantum Zeno effect. Whereas, in the light and sunglasses experiment, we rotated the angle of polarization of light by quantum measurements performed by a series of polaroid lenses; in the quantum proto-cell, it was a series of proto-enzymes that performed the quantum measurements. In the light experiment, it was our observation of light passing through the horizontal lens that performed the final irreversible act of measurement which amplified the quantum system; but in the proto-cell, it was the emergence of the self-replicator that nailed the system to the classical world. The emergence of the self-replicator was the end of the line for the inverse quantum Zeno effect. Quantum measurements by proto-enzymes along the route to the self-replicator laid down a series of stepping stones that led to the emergence of life.

Only a few measurement steps would have been necessary to enhance significantly enhance the probability of generating the first self-replicator. Remember that a single Polaroid lens performing quantum measurement can increase the probability for photon transmission through a crossed lens system (combination of horizontal and vertical lenses) from zero to six per cent – an infinite increase. Even if only two or three steps along the path to self-replication were subject to quantum measurement, it may have been sufficient to raise the probability of our self-replicating peptide from $\frac{1}{10^{32}}$, to a value achievable in a quantum proto-cell within a small warm primordial pond. In this guise, the inverse quantum Zeno effect performed a role later taken over by Darwinian natural selection. Just as the evolution of the highly improbable structure of the eye was guided through the multiverse of design by natural selection's ability to *capture* beneficial mutations, so the evolution of the highly improbable self-replicator was guided through a prebiotic chemical multiverse by quantum measurement's ability to capture the quantum states which led to the self-replicator.

WHAT HAPPENED NEXT?

Our model thirty-two-amino-acid peptide (or any other kind of self-replicator) still had a very long way to go before it reached the complexity of the bacterial fossils found at Chinaman Creek. But once self-replication was established, Darwinian natural selection could have taken over. The replication machinery of the earliest self-replicator would undoubtedly have been prone to error, allowing the generation of many mutant molecules. Most would probably have been less efficient at replicating than their parent structures. However, a few would have been even more successful and Darwinian natural selection, acting at a molecular level, would have increased the proportion of these second-generation replicators in the primordial soup. Further rounds of replication, mutation and molecular evolution would have led to a gradual increase in self-replicating *fitness*. The self-replicators must at some stage have captured lipid membranes, peptides or nucleic acids. These would have protected the self-replicators, and helped in their replication. Eventually the first living cell would have emerged.

However, the problem with this standard molecular evolutionary scenario is that it is very hard to drive this kind of chemical evolution in the direction of increasing complexity. Many researchers have tried to mimic this step in the laboratory, but with only limited success. These experiments are usually performed by using enzymes to copy DNA or RNA molecules. A popular system is provided by a virus called Qβ that uses the enzyme *Qβ replicase* to replicate its RNA chromosome. The enzyme can be extracted from the virus particles and used to replicate any RNA template. Its replication has quite a high error rate, so it generates many mutant RNA templates. Some of these are more efficiently copied than the original template, some less so. Competition between variant RNA templates does indeed drive molecular evolution. The starting RNA is soon out-competed by mutant RNA species better at being replicated. However, supply the system with any complex RNA molecule and the progeny produced after many hundreds of cycles of replication are invariably smaller, simpler RNA molecules. The system never evolves in the opposite direction – towards greater complexity.

A number of other biochemical systems have been used to mimic molecular evolution, but they all suffer the same fate: simple systems get simpler – the opposite of biological evolution. Of course, these laboratory simulations of molecular evolution are still far from real prebiotic molecular evolution. For one thing, they still rely upon enzymes extracted from living cells for their replication. Artificial Life (Alife) researchers who abandon wet biology for *in silico* 'life' take an entirely different approach. Alife researchers use computer programs called *genetic algorithms* to simulate life. This is not as strange as it may seem as DNA is, of course, a code rather like a computer code, only it is our cells rather than a computer that reads the code. One of the most fascinating Alife systems was developed by the ecologist-turned-Alife-researcher, Thomas Ray of Delaware University. Ray invented a computer language called *Tierra* that forms a kind of digital primordial soup. He then constructed *Tierrans* – simple programs – that self-replicated within the Tierra digital world. Ray ensured the replication machinery was imperfect, so that mutant Tierrans, with errors in their instruction set, were generated alongside perfect copies. The programs were seeded into a Tierran world where they would all compete for resources – essentially computer memory – allowing a digital version of Darwinian natural selection to drive Tierran evolution.

Ray seeded his first Tierra world with the simplest program he could devise – just thirty-two lines of instruction – which was capable of self-replication. However, when released into his computer, it quickly generated mutants, many with fewer lines of code than their parent, but still capable of self-replication. Soon a whole ecosystem sprang up of self-replicating Tierrans, together with various forms of parasitic computer code. Fascinating though this digital evolution and ecology certainly was, what Ray did not observe was any drive towards increasing complexity. His digital organisms (d)evolved relentlessly towards greater simplicity and efficiency rather than evolving towards the increasing complexity characteristic of biological ecosystems.

Copies of Tierra have been run on thousands of computers across the world to generate billions of variant Tierrans, but so far nothing more interesting than Ray's first simulation has emerged. Many other digital creatures have been constructed by other Alife researchers (including one called CREATURES constructed by Greg Knowles and myself[4]) but they all evolve in the same direction – towards decreasing complexity.

So laboratory simulations of molecular evolution, either *wet* or digital, are beset by the same problem: an inexorable drive towards simplicity. Yet biological evolution has moved in precisely the opposite direction, towards increasing complexity. Some ingredient must be missing from molecular evolution experiments. I believe that the missing ingredient is quantum evolution. Let's go back to our earliest primordial self-replicators *living* within its quantum proto-cell. Remember that we have modelled self-replication on David Lee's thirty-two-amino-acid self-replicating peptide. This molecule has very specific *food* needs – it must have its seventeen amino acid and fifteen amino acid progenitor molecules to stitch together enzymatically to make copies of itself. However, it is likely that the supplies of these molecules (or similar progenitor molecules) would have been very limited in the primordial soup. Our self-replicator would have soon exhausted them all. Without substrates, the self-replicator would no longer have been able to act as an enzyme.

Once the self-replicator ran out of substrates, its ability to perform quantum measurement upon itself would have evaporated. Without quantum measurement, the self-replicator would inevitably have re-entered the quantum realm. Further chemical modifications (additions of further amino acids) to the self-replicators would have taken place within the quantum multiverse to generate a superposition of modified and yet more complex self-replicators. Whenever these modifications led to the emergence of new enzymatic activities, measurement would have collapsed their quantum states. However, once again the emerging proto-enzymes would have emerged unscathed from the measurement, and been free to re-enter the quantum world. But if the amino acid additions led to the construction of a new self-replicator that could replicate in the substrate-depleted environment, their quantum state would have inevitably been amplified to the classical level. The new self-replicator would once again nail the quantum system to the classical world. Again we are left with a chain of quantum measurement leading to self-replication and concomitant irreversible amplification to the classical level: the inverse quantum Zeno effect. New structures emerged that had the same skills as the original self-replicator, but with additional abilities to utilize novel substrates for self-replication – perhaps stitching together three or four smaller (and more numerous) peptide fragments to make a copy of the self-replicator. In this guise, quantum evolution would have continually driven the system towards increasing interaction with the environment, thereby increasing complexity.

Additional biochemicals may have been recruited – RNA, DNA, peptides, lipids and sugars – which increased the efficiency of replication. However, an essential component necessary to ensure the continuance of quantum evolution was the protection of the genetic material (whether RNA, DNA, peptide or some completely novel molecule) inside some kind of proto-cell, to maintain quantum coherence. Initially, the self-replicators would have invaded any suitable structures, such as pores in rocks or inorganic microvesicles. However, mutant self-replicators that learned to construct their own proto-cells – which protected the genetic material from the vagaries of the external environment – would have been more successful at quantum evolution and thus left more descendants. Perhaps the first cellular life was some simple self-replicator, sheltering within nanometre-scale microspheres, not very different from the 'nanobacteria' found in terrestrial subterranean rocks, or the fossil-like structures within the ALH84001 Martian meteorite. Eventually, the proto-cell became entirely self-supporting and the first true living cell, our first common ancestor, was born.

LIFE IN THE LABORATORY?

The question that might now be running through your head (it has certainly run through mine) is: *if life is this easy, why hasn't anybody made it in the laboratory?* The answer may be to do with the problems of bucket chemistry alluded to above. The inverse quantum Zeno effect depends on the ability of single quantum states to be decomposed into a superposition of two or more orthogonal states. But primordial soup experiments are performed with trillions of molecules in typical thermodynamic conditions where quantum superpositions would quickly evaporate under decoherence's influence. Zeno can no longer help us.

One solution may be to perform primordial chemistry in situations that promote the retention of quantum level phenomena, such as the carbon nanotubes mentioned previously. Perhaps when we know more about how quantum coherence is maintained in high-temperature superconductors, new approaches to stimulating quantum chemistry in the laboratory may be discovered. We may then be able to synthesize life in the laboratory.

An entirely different approach could be to stimulate quantum evolution in a computer. The digital Alife creatures (like the Tierrans) presently function entirely at the classical level. Their self-replication machinery is a series of binary switches that involve the movement of millions of electrons across electrical junctions. They are incapable of quantum evolution and so only evolve *downwards* towards greater simplicity. However, switching devices have recently been constructed that respond to the passage of single electrons and thereby obey quantum laws. Scientists are aiming to use these devices to construct *quantum computers* capable of executing massively parallel computation as a quantum superposition. This concept is analogous in many ways to quantum evolution, where living systems explore all possible interactions with their environment as a quantum superposition. No quantum computer has yet been constructed by man but life may have discovered quantum computation more than three and a half billion years ago. When quantum computers are used as the seed organisms for Alife, (qAlife?), they may explore the quantum multiverse of digital evolution, just as the earliest proto-enzymes and self-replicators explored the quantum multiverse of chemical evolution. qAlife creatures will then be driven, by quantum measurement, towards increasing interactions with their digital environment and increasing complexity. Digital life will then become a reality.

So quantum measurement may be the key ingredient missing in conventional theories to account for the origin of life. The ability of self-replicators to amplify their own quantum state to the classical realm, and thereby measure that state, was crucial to life's emergence. Once the first self-replicating cells were generated, the rest, as they say, is history. However, quantum evolution may not have played out its role with the emergence of life. Indeed, I believe it is still going on today. To find out why, we must next explore the product of those early evolutionary events, the living cell.

11

The Quantum Cell

In Chapter Five, we explored how all the actions performed by living cells involve proton and electron movement; discovering later that the motion of these particles is governed by the strange rules of quantum mechanics. Yet, despite being made of these quantum particles, I have never seen a cat that could walk both ways round a block simultaneously. There must be a border between the quantum and the classical world: one that lies somewhere between the protons that make up the cat, and the cat. A glance down a microscope should convince you that whole living cells, like cats, inhabit the classical world. No muscle cell can both contract and relax at the same time. No blood cell is able to pass through two blood vessels simultaneously. The border must lie somewhere above the level of electrons and protons but below the level of a whole cell. Where is it?

To help us find where the quantum-classical border lies in living systems, let us first try to locate it in the inanimate world.

WHERE IS THE QUANTUM-CLASSICAL BORDER IN DEAD CATS?

Living creatures are made of exactly the same materials as inanimate objects, so a knowledge of where the border lies in non-living systems (much easier to study than living ones) should give us some pointers where to look inside living cells. Imagine first trapping a proton in a box held at room temperature, and then slowly cooling the box. At room temperature, our warm proton possesses a considerable quantity of thermal (kinetic) energy and zips speedily from one end of the box to the other, bouncing against its walls and exchanging kinetic energy with the box. As long as the box is held at room temperature, the collisions will exchange an equal amount of energy and the proton will keep the same thermal energy. However, as the box gets colder, its walls hold less thermal energy, so with each collision our bouncing proton loses a small quantity of kinetic energy to the box. As the box cools further, the bouncing proton slows with each collision until at −273°C (absolute zero on the Kelvin temperature scale) our proton slows to a standstill.[1] Zero degrees Kelvin is the temperature at which all random molecular motion due to heat ceases. At zero degrees, heat no longer exists. Yet despite being at rest, our proton will not be entirely still. Even at zero degrees, it will possess an intrinsic vibration, a consequence of its *ground state energy* and its dual nature as both particle and wave (remember $E=hv$, from Chapter Seven). The proton will therefore vibrate with a single pure frequency (which can be calculated from the knowledge of its energy) and behave as a *quantum harmonic oscillator*.

As a quantum oscillator, it can be considered as a kind of quantum tuning fork, whose wave pattern fills the entire box. An alternative view is of the proton as a particle suspended in a superposition of all possible positions inside the box.[2] If the box contains regions of differing potential energy (perhaps there is a slight charge difference across the box), the particle will also exist as a superposition of energy states. Particle or wave? Both viewpoints are equally valid.

What happens if we warm our proton to body temperature? Will it lose its quantum character and occupy a real position inside the box? Absolutely not! Low temperatures are not necessary for quantum behaviour of single particles. Warming the box increases the proton's thermal energy but its wave function continues to fill the entire box. Its wavelength merely shortens, in a series of discrete quantum jumps, to accommodate its higher energy (again from $E=hv$); but the proton (viewed as a particle) continues to exist as a quantum superposition. This lack of temperature dependence of simple quantum systems is important to bear in mind when we come to look at living cells. It is often stated that quantum effects are limited to ultra-low temperatures but this is only true for phase-dependent phenomena, such as interference occurring between large numbers of particles. Phenomena such as quantum jumps, quantum superposition, uncertainty and measurement-influenced dynamics are all retained for simple quantum systems, even at high temperatures. They are just harder to detect.

What if we add a second proton to the box still at absolute zero? The second proton will exist, the first, as a superposition of position and energy states. Because all protons have the same ground state energy and are indistinguishable, quantum laws prevent them from vibrating out of phase (since this would make them distinguishable) so they vibrate coherently (in phase) as a single quantum harmonic oscillator.

Now what happens if we warm the box? Initially, the protons continue to vibrate in phase and remain coherent but the greater thermal agitation makes it increasingly likely that they will drift out of phase and decohere. How hot does the box have to get to make the protons decohere? Wojciech H. Zurek of New Mexico's Los Alamos National Laboratory has derived an equation that can be used to estimate *decoherence time* for simple quantum systems.[3] This depends on a number of factors. The first is mass. As all teachers know, it is far harder to control a big class of rowdy youngsters than a small one. Coherent motion is similarly far easier to maintain for small number of particles than for large. So increased mass tends to reduce decoherence times. The second factor is the displacement of the quantum states. If alternative particle positions are far apart then they tend to decohere faster. The third factor is temperature: as atoms and molecules get hotter, their random jiggling tends to upset their phase relationships; decoherence times get shorter as temperature increases. The fourth factor is the system's flexibility. Rigid systems maintain coherent vibrations for much longer – which is why tuning forks are made of metal rather than rubber.

This last factor – flexibility of the quantum system – is likely to be important in biological systems, but unfortunately, it is the hardest to evaluate. It depends on all sorts of factors to do with the energy barriers between a system's different states. These are very difficult to estimate for complex biomolecules stuck inside the cytoplasm of a living cell.

Although the equations are rather crude and, even for simple systems, many of the parameters can only be estimated, they have provided decoherence rate predictions that have stood up well to experimental tests in laboratories across the world. Protons or electrons displaced by atomic-scale distances are predicted to remain coherent for very long times – even at room temperature. Big objects (like cats) have far smaller decoherence time, mainly because of their increased mass. In fact, it is the square of mass which appears as the denominator in Zurek's equation. Increasing the mass of a quantum system tenfold, decreases decoherence time a hundredfold (10^2); increasing mass one millionfold, decreases decoherence time a trillionfold. It is therefore hardly surprising that quantum coherence effects are not apparent in big objects. For a cat with about 10^{30} molecules, the decoherence time will be less than a trillionth of a millisecond.

Decoherence estimates do, however, have their limitations. In particular, Zurek's equation can only be applied successfully to very simple systems. We know some complex materials can retain quantum coherence at relatively high temperatures, most notably superconductors. But estimates of decoherence times allow us to mark out a rough border between the quantum and classical worlds. However, this border only strictly applies to phase-dependent phenomena. Quantum jumps, uncertainty and even quantum superpositions may persist beyond this border but the evidence will be harder to find. Another important point to bear in mind when we look at living cells, is that decoherence does not *go all the way down*. What I mean is that although quantum-level phenomena are not visible in a (hot) object the size of a cat, the individual electrons and protons inside will continue to display quantum-level phenomena at any temperature. If we could examine a pair of protons inside a cat we would find the same interference phenomena as the pair of protons inside the empty box. It is only in considering the whole bulk of a big object that the interference terms get washed out.[4] So quantum effects do not disappear from big things in the world – they just become harder to find.

Quantum effects are not of course limited to biological phenomena. Electrical engineers have to deal with quantum-level effects (particularly quantum interference) when they construct nanoscale microelectronic devices for new generation computers. But living cells are already working at these levels. The mitochondria that transport electrons and protons are less than one micron (a millionth of a metre) in diameter and their membranes (where most of the action takes place) have a width of about only forty nanometres (forty billionths of a metre). The cross-sectional area of the DNA double helix is only two nanometres across (about the same as a C^{60} fullerene molecule which shows quantum effects at temperatures as high as 600°C) and even big protein subunits are usually no bigger than about twenty nanometres. These structures are the *electrical* components of living cells. Electrical engineers have had to deal with quantum-level phenomena only in recent years, but the living cell – a nanoscale device – has been confronting these phenomena for billions of years. Like electrical engineers, the living cell must have learnt to deal with quantum-level effects, and, to exploit quantum mechanics for its own ends.

This chapter will explore how cells may exploit quantum-level phenomena to perform directed actions. But first, we need to discover the limits of quantum mechanics in living cells.

VOYAGE TO THE BOTTOM OF THE CELL

The border between the quantum and the classical world must lie somewhere below the level of whole cells. To find it, we must look inside living cells. Bacteria have the simplest cells, so they are the easiest place to start. Most are very small: about a thousandth of a millimetre long, far too small to be seen by the naked eye. In fact a million cells of the common gut bacterium *E. coli*[5] could be squeezed into a sphere with the diameter of a full stop. Even with a powerful light microscope, the *E. coli* cell appears as a short featureless rod. Although we could examine the cell interior with a powerful electron microscope and see structures, the cell would first have to be frozen, sliced, dried and, inevitably, killed. We need to examine a living cell *in action*.

A variety of techniques can be used to observe a live cell, but none gives the resolution we need to explore quantum-level events. Instead, we will borrow a device from the classic 1966 film, *Fantastic Voyage*, where a crew of scientists aboard a tiny submarine are miniaturized and injected into the bloodstream of a dying man to perform delicate brain surgery. Let us imagine ourselves shrunk similarly to a size of just a few tens of nanometres – much smaller than the *E. coli* cell – and swallowed by a human host. Just like the movie's heroes, our fantastic voyage will be inside a miniaturized submarine and, after a thrilling ride through the stomach and ileum, the craft locates our prey: a torpedo-shaped *E. coli* cell swimming through the large intestine. The bacterium is driving itself through the gut, powered by a rotating bundle of protein rods (the flagella) at the cell's rear end, that acts as a kind of outboard motor. Synchronizing our velocity with the speeding cell, we lower the vessel down onto its surface.

We find ourselves parked on a thick waxy membrane studded with protein spikes and perforated with ring-shaped portals. These protein-lined pores act both as mouths and as drains for the cell, pumping nutrients in and expelling waste products. We dive into one drain, which takes us through the outer membrane and into a water-filled chamber lying between it and an inner membrane. The same protein-lined shafts pierce this internal membrane, so we dart into one of them and emerge inside the bacterial cell's interior.

The first thing we notice is that we have to drive our engines a lot harder to move through the cell inside. This is because water inside cells – in the *cytoplasm* – is not quite liquid. Much is tightly bound to big bulky proteins. What little free water there is, is highly structured: not solid but not quite water either, perhaps resembling jelly. Hanging within the cytoplasm is the cell's machinery: ribosomes which churn out ribbons of proteins; docking stations which ferry some proteins through the membranes and out of the cell; rotary engines which power the cell's spinning flagella; and, anchored beneath the inner membrane, the proton-powered turbine engines which make ATP. At the cell's core lies its nerve centre: a coil of circular DNA. Most of the cell's single chromosome lies dormant – densely coiled and wrapped in protein – but loops of active DNA are busily directing the construction of strings of messenger RNA which ferry the genetic code to the ribosomes. One DNA loop is being dragged through a ring of proteins attached to the inner membrane. As it is fed through, the double helix is pulled apart and a new pair of DNA strands manufactured, using the old strands as templates.

The first target in our search for the quantum-classical border inside the cell is a proton, part of one of the cell's many proteins: a single molecule of the enzyme called *beta-galactosidase*. The enzyme looks rather like any other protein inside the cell: a tightly knotted bundle of amino acid rope made up of about a thousand amino acids. But this enzyme is currently inactive. Its job is to hydrolyse (react with water) the disaccharide milk sugar lactose, breaking into its component bits: glucose and galactose. However, the human host to our *E. coli* cell has not drunk any milk since breakfast and it is now the middle of the night. The enzyme has nothing to do until the next lactose arrives, along with the breakfast cereal, the next morning. Our cell has been without food for some time now and has exhausted its reserves. To conserve energy, it switches itself into a kind of hibernation state (dormancy), until the lactose arrives.

Our information is that within the beta-galactosidase enzyme lies our target proton on one of the protein's amino acids. This proton (a hydrogen nucleus) is attached to an oxygen atom within the amino acid molecule, by a covalent bond. Our sources also tell us that nearby lies a nitrogen atom which, like the oxygen atom, is relatively electron rich and would *like* to capture our target proton. We imagine that, if supplied with enough energy, then our proton might escape the pull of the oxygen atom's electrons and hop onto the nitrogen atom. In fact, let us suppose that calculations indicate that at body temperature the surrounding thermal energy gives the proton a fifty per cent chance of hopping from one atom to another. If, at some later time, we were to use some ultra-powerful electron microscope to locate the proton, we would find it still attached to the oxygen atom about fifty per cent of the times we looked, and attached to the nitrogen atoms for the remainder. Our task is then to discover: does the proton inhabit the quantum or classical realm? Or, to put the question another way: is the proton's position real?

It is absolutely clear that our target proton is a quantum particle that will exist in a quantum superposition of states until measured. Avoiding the Copenhagenist interpretation that our presence as observers is necessary to collapse the wave function describing the particle, we will instead look for evidence of our target proton interacting with a complex environment to cause decoherence and thereby *observer-free* measurement. But what level of interaction will suffice to cause the decoherence of protons inside living cells? Does our target proton live above or below the quantum/classical border?

THE QUANTUM BORDERLAND

In *What is Life?* Schrödinger postulated: 'The living organism seems to be a macroscopic system which in part of its behaviour approaches purely mechanical (as contrasted to thermodynamical) behaviour to which all systems tend, as the temperature approaches the absolute zero and the molecular disorder is removed.' As discussed above, absolute zero is the temperature at which decoherence is completely suppressed and all matter falls under the sway of quantum rules. Schrödinger was therefore suggesting that decoherence is somehow suppressed inside living organisms, allowing them to obey quantum rules at high temperatures. Sadly, we cannot rely upon such an eminent authority as Schrödinger to find the quantum/classical border inside living cells. We must instead look towards estimates of decoherence times.

Unfortunately, we cannot use Zurek's equation directly to come up with a decoherence time for a proton inside a living cell. There are too many unknowns. We can however use it to give us a few pointers towards those factors likely to be important. Clearly mass and temperature will work against quantum coherence. A typical protein has a mass many thousands of times bigger than a single proton, so its decoherence time will be correspondingly reduced. On the other hand, displacement distances will be small within a protein, so this will tend to work towards maintaining coherence. Flexibility is also very restricted in biological systems due to the high density of electrostatic forces. Some measure of the limited freedom of intracellular protons can be obtained by a technique called *nuclear magnetic resonance* (NMR), which moves a proton in a magnetic field and then allows it to relax. The time it takes for the proton to fall back into its original state gives a rough measure of proton flexibility. The protons inside living cells take from milliseconds to tens of seconds to relax – indicating that the intracellular matrix is more rigid than might be expected from its chemical composition alone. And proton NMR measures only the relaxation time for the majority of protons attached to water molecules inside cells. Protons buried within biomolecules such as proteins or DNA are likely to have much longer (though harder to measure) relaxation times and correspondingly lengthy decoherence times.

Another factor that will interact with the biological molecules' reduced flexibility is the energy flux through the system. The physicist Herbert Frölich proposed that if metabolic energy were supplied to biological molecules inside living cells at a high enough rate, they might be forced to oscillate at a single coherent frequency. This kind of metabolic energy pumping is analogous to the energy pumping which gives rise to pulses of coherent photons inside lasers. There is no convincing evidence as yet for these high-frequency coherent oscillations inside living cells, but the possibility of energy pumping does emphasize the considerable uncertainty in our understanding of the quantum dynamics inside living cells. There have been a number of reports of quantum-coherent phenomena in biological systems, suggesting that decoherence times may be lengthy. S. Gider from the University of California detected quantum magnetic phenomena in the iron-carrying protein called ferritin. Electron tunnelling (which is of course a quantum phenomenon) is widely assumed to play a role in electron transport in respiration, photosynthesis and in many enzyme mechanisms.[6] Weak electromagnetic fields have been shown to have surprising effects on biological systems – from making nematode worms grow faster to changing genes expression levels or preventing cell apoptosis (a kind of cell suicide). It is extremely hard to account for how these low-strength fields exert their influence by any conventional electromagnetic induction effects. However, phase coherence is exquisitely sensitive to electromagnetic perturbations and it is therefore possible that it is coherent oscillations inside living cells that are the targets of these low-strength fields.

You may well remember from Chapter Nine that a useful way to consider decoherence is as a measure of the leakage of information concerning a quantum system's state into a complex environment. In the two-slit experiment it was 'which way?' information that leaked into the environment to cause decoherence. For our protons, it is which way has the proton gone: to the oxygen or nitrogen atom?

Imagine first that the position of our target proton has no significance to the protein or its surrounding cell (there are likely to be very many protons in this situation inside living cells). The proton will be buried within the globular structure of the protein molecule in some out of the way stretch of the amino acid chain, playing no role in the enzyme's catalytic activity. Although the proton shift may cause changes in the enzyme's structure, due to the shift in our target proton's position and the subsequent reshaping of its electric field, these are likely to be very small, insufficient on their own to cause decoherence. The proton's position will not become entangled with the external environment. We could stare as long as we like at the target protein from our miniaturized submarine's cockpit but will not spot any information that correlates with the particle's position in the cell environment. With no information leakage, decoherence will be retarded and the proton will remain indefinitely as a quantum superposition.

But the position of many protons in the cell will not be so inconsequential. Remember that enzymes are proteins which catalyse reactions by mobilizing the electrons and protons in their substrates. They do so by *attacking* the substrates with their own particles: the electrons and protons attached to their amino acids. So the precise location of their electrons and protons is crucial for their enzymatic activity. Now imagine that our target proton is bound to an oxygen atom on the 461st amino acid of the beta-galactosidase enzyme: a glutamine. This proton is particularly critical for enzyme activity, since lactose breakdown is initiated by its being fired into the heart of the lactose molecule, where it destabilizes chemical bonds. If the proton has however gone over to the adjacent nitrogen atom, then it cannot be fired into the substrate and the enzyme will not work. So the enzyme can *measure* its own proton's position, if lactose is available as a substrate.

But how will we *see* the breakdown of lactose? Fortunately, it will be easy, since lactose breakdown does not stop with glucose and galactose generation. The products are drawn instead into a complex web of metabolic pathways, where they will be broken down further and their electrons harvested and fed into the respiratory chain that makes ATP (as described in Chapter Five). The ATP energy may then be utilized in various ways by the cell. It may *wake* the cell out of its dormancy. It may be used to provide energy to power the flagella motor in the cell. The energy could be used for cell division or equally to supply the power for biosynthesis of new proteins. Each possibility will cause massive changes to the positions and energies of billions of particles both inside and outside the cell which are easily spottable. If we witnessed any of these classical level events, we could be sure the target proton was attached to the oxygen atom. Equally, their absence would constitute a null measurement of the particle's position; in which case we would know that it is attached to the nitrogen atom.

In fact we don't even need to be miniaturized to witness these classical signals. A full-sized microscopist could detect the cell's motility or a molecular biologist could detect its DNA replication. The host to our *E. coli* cell could even suffer a bout of urgently classical diarrhoea from an infection initiated by the revived bacterium. Just as a Geiger counter can generate a classical signal that betrays whether or not a radioactive atom has decayed, so a living cell can generate a classical signal that betrays proton position. The living cell is turned in upon itself to perform measurements on its own particles and thereby *internal quantum measurement*.

An important point to note is that our target proton will only be measurable when lactose is available in the cell. If there is no lactose, the proton position will make no difference to the cell and no information correlating with particle position will leak into the surrounding cell. However hard we stare at the cell, we will be unable to discern whether the proton is attached to the oxygen or nitrogen atom. Under these circumstances, decoherence will be suppressed and the proton will remain as a quantum superposition. Quantum measurement will be *conditional* upon the state of the cell.

There is of course nothing special about beta-galactosidase enzyme's role as the target of internal quantum measurement. The average bacterial cell is capable of producing more than a thousand different proteins and each will have fundamental particles poised within their active sites, subject to quantum measurement by the cell. For our target enzyme, internal quantum measurement was conditional upon lactose presence in the cell. For other components, it might be different substrates, temperature, light, brightness or any other physical parameter that could affect the cell's ability to perform internal quantum measurement. Measurement will also not be confined to particle position. Real values for energy, momentum, spin, or indeed any quantum property, will be similarly conditional upon quantum measurement. The chain of entanglement between the particles inside cells and their environment ensures that living cells are uniquely sensitive to all kinds of quantum events taking place inside them.

Our mission was to find the border between the quantum and classical realm inside living cells. What we have discovered is that there is no fixed border inside living cells but rather one that shifts up and down the cell function hierarchy, depending on the state of the cell and the resources available. In starved, inactive cells, most fundamental particles will be sunk into the quantum world of superpositions and interference. But feed the cell and its quantum-measuring apparatus will be armed with substrates that allow it to perform densely spaced measurements on critical particles, forcing them to take on real values and inhabit the classical world. The quantum-classical border will thereby be pushed down into the bowels of the cell, where only non-critical particles, shielded from environmental interactions, will continue as quantum entities.

THE QUANTUM CELL

The classical view of the dynamics inside living cells (still the view held by most biologists) was of classical particles pursuing independent trajectories through intracellular spaces. This vision allowed biochemists and geneticists to wholeheartedly adopt the reductionist programme of dissecting the cell into smaller and smaller pieces, with the expectation of gaining a greater and greater level of understanding. However, now biology has reached the level of fundamental particles, we must confront the quantum cell, which has revealed itself as a dynamic mosaic of quantum and classical states. Particles can no longer be considered as independent entities but as the products of internal quantum measurement. Quantum mechanics directs us to look up from the fundamental particles and examine the environment measuring them.

Why does this matter? The earlier chapters on quantum mechanics showed that quantum measurement has striking effects on the systems it measures. Measurement of quantum particles is never innocuous; it always affects dynamics. Physicists are normally employed to make quantum measurements, and the choices they make (concerning the properties they wish to measure) affect the dynamics of the systems they study. But now, we have the living cell as an independent quantum-measuring device that measures its own state, so that the *choices* it makes about what it *wishes* to measure will influence its internal dynamics.

I must immediately qualify this by saying that I do not believe that *E. coli* cells or even individual animal cells have any volition over their actions. The environment of the cell arms their quantum-measuring devices and thereby determines the properties that the cell can measure. This will in turn influence the internal dynamics of the cell. This represents a kind of choice, since it is an influence denied to inanimate objects unable to measure the quantum states of the particles within them. But – at least in simple living systems – the ability to make these choices is not associated with any *conscious* decision. Nevertheless, I do believe that this ability to make quantum choices is the basis for *our* sense of volition as conscious beings. But more about that in the final chapter. We must first investigate the consequences of quantum choice for all living cells.

There are many ways that quantum measurements could influence particle dynamics inside living cells. The quantum Zeno effect and its inverse provide particularly interesting possibilities. In the last chapter we looked at how quantum measurement by enzymes may have precipitated the emergence of the first self-replicators. But quantum measurement by enzymes is taking place within all our cells every day. What are the consequences on the dynamics of the particles inside our cells?

Consider a simple action which may be performed or not performed on the basis of a chain of possible quantum events; perhaps the reactions required to fire the motor that drives the cell's flagella in response to a signal that there is food in the vicinity. We will imagine that firing-up the flagella requires many thousands of ATP molecules, rather more than the sixty or so delivered by a single round of lactose hydrolysis. We will therefore need our beta-galactosidase enzyme to hydrolyse as many as a hundred ATP molecules to initiate the action. For each hydrolysis reaction, our target proton in the active site of the enzyme must be attached to the oxygen atom. If the proton has skipped over to the nitrogen atom, the lactose will not be hydrolysed. There is a fifty per cent chance that each molecule of lactose will be hydrolysed when it encounters the enzyme (whenever it finds the proton attached to the oxygen atom).[7] Although, if given enough time, our cell will eventually hydrolyse sufficient lactose to power the flagella, we will add more drama by imagining that there is a race to get to the food first. Another bacterium is already bearing down upon the potential meal and if our cell doesn't manage to hydrolyse all one hundred lactose molecules in one uninterrupted sequence and get off to a quick start, then it will arrive too late. The probability of obtaining an entire interrupted sequence of one hundred molecules of lactose being hydrolysed will be 2^{100} or approximately 10^{30}.

Considered classically, it is highly unlikely that the cell will perform the required action, reaching the food first. However, this brings us back to our stroll through a multi-dimensional landscape. This time the landscape has only two dimensions, corresponding to proton at oxygen and proton at nitrogen. The route to successful rapid firing of the flagella motor represents just one very fine path (a hundred steps of *proton at oxygen*) through this binary quantum landscape. And, just as we considered the events leading to the origin of life in the last chapter, each step may be subject to quantum measurement. In this case it is easy to see what could perform the measurements: the beta-galactosidase enzyme. Each hydrolysis of a lactose molecule will potentially perform a measurement of the proton's position. However, in most cases, the enzyme may emerge unscathed from the quantum measurement and be free to re-enter the quantum world. Yet, once sufficient lactose has been hydrolysed to potentially fire the cell's flagella, then the quantum state of the cell will inevitably be amplified to the classical level. The cell will (if the correct sequence of quantum events has taken place) fire up the flagella, drive itself through the gut and irreversibly collapse the quantum state of the enzyme to the classical level.

So once again we have a series of quantum measurements leading to an irreversible amplification to the classical realm. Once more, the inverse quantum Zeno effect can enhance the route's probability of leading to the final measurement. Just as a dense series of measurements (by polaroid lenses) can raise the probability of light travelling along a path of changing polarization angles to reach a detector, so a dense series of (enzymatic) measurements along the path towards firing-up the cell's flagella motor, may *capture* the cell along that path and increase that action's probability. The firing of the cell's flagella will represent the final amplification step – the end of the line for the inverse quantum Zeno effect – that will collapse the quantum system to a classical state. The cell may propel itself forward, by quantum measurement, to get to the food first. It will perform a *directed action*.

Although this scenario may appear somewhat contrived, in principle it is not at all unlikely. Bacteria, and indeed nearly all living creatures, live at the very edge of subsistence where single quantum events or a series of quantum events can make a very real impact on their chances of survival. For most creatures, it will not be the enzyme beta-galactosidase that will be so critically poised, but there are many other enzymes and physical states crucial to their being able to successfully perform a critical action. These actions will be subject to quantum measurement. The densest measurements will be those that involve the greatest number of interactions with the cell's environment. The cell will thereby *direct* itself towards those activities that interact strongly with their environment and perform directed actions.

I should emphasize that I am suggesting an *increase in probability* for actions, not necessarily a way of making those actions happen with one hundred per cent probability. How much quantum measurement can enhance the probability of quantum events inside living cells will depend on the density of the measurement steps, the robustness of quantum superposition states (to decoherence) prior to measurement, and how easily the steps can be decomposed into orthogonal states. Quantum measurement does, however, provide us with a way to give living organisms *an edge* in their interaction with the outside world. Internal quantum measurement confers on living cells an ability to influence their internal particle dynamics in a way unique to life. This influence is key to understanding how living organisms escape the straitjacket of classical determinism. The behaviour of a living organism is poised upon the quantum dynamics of its interior, allowing quantum measurement to provide the critical nudge that tips the organism one way or another to *make choices* and perform directed actions.

WHAT IT MEANS TO BE ALIVE

This view of the cell as a unique quantum-measuring device, capable of *capturing* internal quantum states, brings us right back to Maxwell's demon. If you remember, the demon's job was to operate a trapdoor, connecting two chambers of a box filled with gas molecules. Without the demon, the distribution of gas molecules would be the same in both chambers. However, if the demon opened the trapdoor, allowing through only molecules in one direction (say left to right) whilst stopping those travelling in the opposite direction (right to left), then he could effect a partition of gas molecules within the box, thereby decreasing the system's entropy. However, the catch is that to perform this feat the demon must detect and measure molecular motion, and the price of his measurement is an accompanying increase in entropy associated with the information gain. The price the demon must pay for his information always outweighs the entropy reduction he achieves. The Second Law of thermodynamics remains aloof from the demon's antics.

However, his activities bear many similarities to the role of quantum measurement inside cells. Just as the activity of the demon *captures* low-entropy states in the box, so quantum measurements of particles inside cells capture their low entropy states. Once again there is a price – and our metabolism is that price. Respiration and fermentation – the energy-yielding reactions inside our cells – represent *entropy burns*, to balance the equation. Our cells excrete the high-entropy waste products of our metabolism such as carbon dioxide, water and heat. This is how the cell pays its entropy dues: to build and maintain its quantum-measuring devices. With these devices the cell is able, like Maxwell's demon, to capture internal low-entropy states and maintain the cell's integrity against the randomizing influence of thermodynamics' Second Law.

The cell's ability to capture low-entropy states by performing internal quantum measurement is, I believe, fundamental to what it means to be alive. You may remember (Chapter One) that there are a number of problems with the standard criteria for defining life. Most lists include self-replication, which relegates hale and hearty animals, such as mules, to the category of the dead. Many cells in our body incapable of dividing, such as nerve cells, would similarly be excluded from the living (making our brains particularly *dead*). Conversely, we have difficulty excluding other self-replicating entities such as computer viruses or even those dreaded chain letters, from the living category. Other items on the lists, such as metabolism, are ill-defined, for they may include phenomena such as fires or automobiles. But with internal quantum measurement included as life's key criterion, then these problems disappear. Both mules and nerve cells are perfectly good at performing internal quantum measurement. A chain letter, fire or computer virus is completely inept. A biological virus, such as influenza, is something of a halfway house, since an isolated virus particle is incapable of performing internal quantum measurement. Any changes that take place within its chromosome remain quantum. However, inject the virus into a human cell and those changes may be amplified to the classical realm. Even then, however, it is really the host cell that does the quantum measuring (by obeying the instructions encoded by the virus); the virus, on its own, cannot be considered truly alive.

Internal quantum measurement is also what we lose when we die. The key to the cell's ability to perform quantum measurement is the chain of entanglement from fundamental particles to the environment of the living cell. Like all quantum coherence phenomena, this chain is very fragile and depends on the cell's structural integrity. Break the chain and random thermodynamic motion will rapidly intercede to destroy quantum coherence. With the loss of quantum coherence, the dead cell will be unable to perform quantum measurement or resist the randomizing influence of the Second Law. It will instead be converted into an entirely classical (inanimate) object. Death is the irreversible loss of quantum measurement within the cell.

So life and internal quantum measurement go side by side. When one is lost, so is the other. Does internal quantum measurement therefore *define life*? Not entirely, for like all 'definitions' of life, internal quantum measurement is not in itself unique to life. *Metastable* devices, such as Geiger counters or atomic bombs, may also detect internal quantum events. They do so by engineering a classical-level phenomenon poised upon the quantum dynamics of their own particles. If the detector of a Geiger counter were pointed towards itself, it too would be capable of measuring quantum events (radioactive decay) taking place inside. Would it be alive? Of course not. Quantum measurement is necessary for life, but not sufficient to define it. We should not be worried about the lack of exclusivity of life's criteria. It has been argued (I think correctly) that no definition of life can be entirely sufficient and necessary; because, if it were, it would imply that life possesses some factor not found at all in the inorganic world. Modern science (including this book) completely rejects the concept of anything *magical* in life; so it should always be possible to find aspects of the living in the non-living world. For a more complete definition of life I would add: *life is a system that uses internal quantum measurement to capture low-entropy states that sustain the state of the system against thermodynamic decay.* It is rather lengthy, but the additional criterion of sustainability excludes atomic bombs (since they cannot sustain their state beyond detonation) and Geiger counters (since their measurements are not utilized to sustain their state against thermodynamic decay) and bring in considerations of life span and self-replication. Mules use quantum measurement to sustain their state for their lifetime's duration. A self-replicating organism can sustain this ability throughout evolutionary timescales.

12

Quantum Evolution

Although quantum measurement may be involved in living creatures' ability to direct their actions, it is difficult to discern its impact, since in most instances it will only provide a slight (often critical) nudge in the direction of a particular action. To see its effects more clearly we need that nudge, or directed action, to be *fixed* in some way. Actions get fixed inside living cells if they cause changes to the cell's heritable material. This is the basis of quantum evolution.

Aboard our miniaturized submarine, let us dive once again into the entrails of another *E. coli* cell. Our target will be the same enzyme we met in the last section, beta-galactosidase, but this time it appears to be malfunctioning. While the protein examined in the last chapter was busy breaking lactose molecules into pieces and kicking the cell into action, this enzyme seems uninterested. Lactose molecules drifting into the protein float right out again unchanged. Yet the enzyme looks the same molecule we saw before – that same tangle of amino acid string. To discover what is wrong, we must descend into the cell's DNA nerve centre and examine the gene that encodes the enzyme.

Located in an unremarkable stretch of about three thousand base-pairs of DNA, the gene appears to be *transcribed* normally into short strings of messenger RNA which ferry the genetic instruction to the ribosomes. The ribosomes appear able to read the message normally and *translate* it into a normal protein. However, when we examine the DNA sequence of the gene (the sequence of *As*, *Ts*, *Gs* and *Cs*), we find what appears to be an error. Where the wild-type gene (the normal functional gene) has the sequence *CTT* encoding glutamine as the 461st codon (the triplet of bases that codes for a particular amino acid), our target gene has instead the

260 · QUANTUM EVOLUTION

sequence TTT.[1] The wild-type CTT DNA sequence, which would have been transcribed by RNA polymerase (the enzyme-making messenger RNA) into the complementary sequence GAA, in messenger RNA, is now transcribed into AAA. Where the wild-type sequence would have been read by the ribosome as a codon for glutamine, the modified AAA messenger RNA sequence will instead instruct the ribosome to insert lysine.

But the glutamine at position 461 is the key amino acid supplying the proton to be fired into the heart of the lactose molecule to initiate its hydrolysis. Substitution with lysine explains the enzyme's inactivity: lysine does not have a proton in the right position to be fired into lactose. Thus the cell cannot feed on lactose and will starve if lactose is the only food available.

But there is a possible way out. You may remember from Chapter Three that one source of errors in DNA replication is the *tautomerization* of DNA bases. Let's concentrate on the incorrect base, T. Normally, when RNA polymerase makes messenger RNA, it will insert a complementary A (adenine) base at this position and this will form part of the AAA codon directing the ribosome to insert lysine into the inactive enzyme. However, if the coding base thymine is in its (rare) *enol* configuration, then this enol thymine could pair instead with guanine (G). The RNA polymerase would insert guanine into the messenger RNA sequence instead of adenine. But this will give us the correct codon: GAA. The messenger RNA could go on to direct the ribosome to make beta-galactosidase with the correct glutamine amino acid at position 461. The resulting enzyme would be active.

So the cell can make a tiny amount of the wild-type enzyme even when its DNA sequence is *wrong*. All that is required is for the coding proton to shift to its tautomeric (quantum tunnelled) position, whilst being used to direct messenger RNA synthesis. However, we have once again slipped back into *classical* talk. A proton is a quantum-mechanical entity that cannot have a defined position in space until a quantum measurement *puts it there*. If there are two possible positions for our target proton on the DNA molecule, it must exist at both as a quantum superposition. Remembering the wave-particle duality of quantum entities, the proton can be considered as a wave (function) that sits above the two possible atomic positions.[2]

When will quantum measurement fix the proton's position? Decoherence will once again provide us with an answer. Decoherence will occur whenever the DNA proton couples with a complex environment. But how does it do that? It is buried within the double helical structure, so the small shift in position when it moves to the tautomeric location will cause only very tiny changes to the particles around it – hardly enough to cause decoherence. But there is another way. When the DNA base – as a quantum superposition – is transcribed into messenger RNA, the messenger RNA will similarly exist as a superposition of states. (This is not so difficult because its coding sequence is made up of the same protons as the DNA code, so they will similarly exist as a superposition of normal and tautomeric position states.) When the RNA arrives at the ribosome, the quantum superposition of coding information will encode a superposition of protein sequence: {beta-galactosidase with lysine at the target amino acid position to make the inactive protein} (+/−) {beta-galactosidase with glutamine at target amino acid position to make the active wild-type protein}. Although the superposition now involves a protein, its alternative amino acid sequences (lysine or glutamine at position 461) will involve only small-scale atomic displacements for a small number of particles. Decoherence may be avoided, at least in the short term, allowing the position state of the DNA coding proton to become entangled with a superposition of active and inactive enzyme. This enzyme superposition may then become entangled with the cell environment.

Will this entanglement precipitate decoherence? This depends whether lactose is available within the cell. If there is no lactose, it will make no difference whether the enzyme is in its inactive or active state. With nothing to do, both enzyme forms will be equally inactive and will remain as a superposition. The DNA proton that encodes the enzyme's alternative states may similarly persist as a superposition. Under these circumstances, there will be a minimal level of entanglement of the coding proton with its environment and so decoherence will be suppressed.

But once lactose enters the cell, the environment will become entangled with the coding proton. Lactose will be broken down by the wild-type enzyme, but untouched by the inactive enzyme. The cell will kick itself into action if the enzyme is active, but remain dormant if it is inactive. The positional state of the DNA target proton will thus become entangled with the positions and energies of lactose molecules, ATP, the cell's state and its surrounding environment. This coupling with a complex environment will cause very rapid decoherence, allowing the cell to *measure* the target proton's position. The cell's ability to measure the position state of the target proton in its chromosome will therefore be conditional upon the lactose availability in the cell.

So our DNA coding proton can become entangled with the cell environment, through its role as the template for messenger RNA synthesis. Another route to quantum entanglement is via DNA replication whilst the proton is in a superposition to generate a daughter chromosome in a superposition of states (again, only small atomic-scale displacements would be generated). One arm of this would encode the parental inactive enzyme, whilst the other arm would encode a *mutant* active enzyme (identical to the wild-type enzyme). Once again, the coding proton will be entangled with the cell's environment and thereby subject to conditional quantum measurement by the cell.

Quantum measurement influences the dynamics of quantum particles. The cell's ability to perform quantum measurements on its coding protons will therefore influence their dynamics. But remember that proton dynamics inside DNA bases are involved in yet another vitally important biological phenomenon: mutation. If the target DNA base is replicated whilst the proton is attached to its normal nitrogen atom, then DNA polymerase will insert the complementary adenine and the daughter DNA molecule will encode exactly the same inactive enzyme as the parent – no mutation. If, however, the DNA base is replicated whilst the proton is attached to the tautomeric (enol) nitrogen atom, then DNA polymerase will insert the incorrect base, guanine, in the daughter DNA strand, thereby generating a mutation.[3] For the proton to be present at either position (rather than in superposition), a quantum measurement needs to be made; and that measurement can be performed only under appropriate environmental conditions (lactose presence). Conditional quantum measurement thereby sits astride the engine of evolution: mutation.

The target gene's frequency of mutation will clearly depend on the proton's dynamics: how much *time*[4] it spends at the tautomeric nitrogen. If this time is short then the mutation will be rare; if it is longer, the mutation will be more frequent. If a dense series of measurements were performed on the target proton at the tautomeric position, then the quantum Zeno effect could freeze its dynamics at that position. But, as we have discovered, the cell itself is able to perform a dense series of measurements on the coding proton if lactose is available. These measurements, in the presence of lactose, may enhance the proton's probability of remaining at the tautomeric position, thereby accelerating the rate of generation of a mutation.

ADAPTIVE MUTATIONS

But environmental enhancement of mutation rates is precisely the phenomenon discovered by John Cairns when he discovered those enigmatic *adaptive mutations* described in Chapter Three. Adaptive mutations occur more frequently when beneficial to the cell, in direct contradiction of the standard neoDarwinian evolutionary theory, which states that mutations always occur randomly with respect to the direction of evolutionary change. John Cairns' initial experiments incubated *E. coli* cells unable to grow on lactose, on media containing lactose, and on parallel media without lactose. If, following standard neoDarwinian evolutionary theory, mutations always occur randomly in relation to the direction of evolutionary change, then the same mutation rate would be expected in both sets of cells. However, Cairns discovered that, after prolonged starvation, mutations that allowed the *E. coli* to utilize lactose increased in frequency. It appeared that the presence of lactose specifically enhanced mutations that allowed the cells to *eat* the lactose. The *E. coli* cell appeared to be able to *direct* its own mutations.

As mentioned in Chapter Three, these experiments are still highly controversial. Strict neoDarwinism is deeply ingrained in current biological thinking: most biologists are very reluctant to accept any revision of its dogmas. There is no question that Cairns' observations are real, but many maintain that there are likely to be more conventional explanations of his experiments than the existence of adaptive mutations. Nevertheless, others such as Professor Barry Hall at Rochester University have also detected adaptive mutations in a variety of bacterial systems. In one of his most recent experiments, Hall measured the mutation rates in non-growing *E. coli* cells for two different DNA bases in the same gene.

When neither gene was beneficial, then mutations occurred at the same rate, but when one conferred a selective advantage, then its mutation rate was enhanced.

The problem with adaptive mutations is to provide a mechanism that could account for them. The interaction between the cell and its environment is conducted at the level of proteins, like beta-galactosidase. The conventional information flow inside living cells is from DNA, to RNA, to protein. There is no conventional path by which information in the cell's environment (lactose presence) can feed back to the DNA that encodes enzymes like beta-galactosidase.[5] The path: gene → messenger RNA → protein → lactose is not reversible. In a recent paper, Professor Hall commented, 'the selective generation of mutations by unknown means is a class of models that cannot, and should not, be rejected.'[6]

Quantum evolution may generate adaptive mutations by providing the required feedback loop: lactose → protein → messenger RNA → gene, via conditional quantum measurement. The living cell's ability to measure the positions of fundamental particles within the DNA double helix will be determined by the composition of its environment in this case, lactose presence. Lactose *arms* the cell's quantum-measuring devices, enabling it to measure the position of the DNA protons that (potentially) encode the beta-galactosidase enzyme. The cell may then perform a dense series of measurements on the position of DNA bases which will perturb the dynamics of those protons, enhancing mutation rates. Quantum measurement may thereby enhance the rate of beneficial mutations to cause adaptive mutations and drive evolution.

THERE IS NOTHING NEW UNDER THE SUN

One of the most thrilling moments experienced by a scientist is when you believe you have come up with a new, exciting way of explaining a phenomenon, previously inexplicable. One of the most disappointing is when someone else beats you to it. Whilst writing this book, I learned that Vasily Ogryzko from Bethesda's NICHD had just published a paper (in 1997) claiming that adaptive mutations are caused by quantum-mechanical measurement effects. Almost simultaneously, Amit Goswami and Dennis Todd of the University of Oregon published a paper in which they also made a case for quantum-mechanical measurement effects behind the phenomenon of adaptive mutations, going on to argue that quantum measurement may similarly be behind the phenomenon of conscious choice.

In case you are wondering why I haven't published my ideas sooner – I tried. Unfortunately, the paper that my colleague Jim Al-Khalili and I submitted in 1996 was rejected as too speculative. Editors of scientific journals are (rightly) very hard to impress when it comes to the theoretical. By this time, I was becoming interested in wider issues concerning the role of quantum mechanics in other aspects of biology. I felt that a book would be a more appropriate vehicle for discussing these more general topics, so I started on the alternative path to this volume, blissfully unaware of any other interest. I was jolted out of my complacency by Vasily's e-mail message. He had seen some Internet postings I had made on quantum effects in biology, and contacted me to inform me about his paper. Like me, Vasily had great trouble publishing his ideas. He gave a presentation of his work at a 1994 Semiotics Congress in Berkeley and, more persistent than I, he was eventually rewarded by his paper's publication in 1997. His success prompted Jim and me to revisit our paper, and we eventually persuaded the American journal *Biosystems* to publish it in 1999.[7]

All ideas have their time and I expect that the twenty-first century will see the flowering of quantum biology. Its roots can be traced back to Schrödinger's (1944) *What is Life*, but though that slim volume stimulated many scientists (such as James Watson) to consider life's physical basis, reductionist biology needed to run its course and dissect living cells down to the level of fundamental particles before its full implications could be appreciated. I believe we are now on the brink of a new adventure which will bring about the synthesis of physical and biological sciences through quantum mechanics. On one hand, electronic engineers are constructing nanotechnology devices – electronics on the scale of living cells – manipulating single atoms and single electrons, on a level where they inevitably confront the quantum nature of their raw materials. Biologists are coming to appreciate the fact that living cells have been performing nanotechnology for billions of years and must have learnt to deal with, and to exploit, the quantum realm. I believe these threads will come together towards a synthesis of disciplines in the new millennium. Electronic devices will be increasingly used to probe the internal workings of the living cell and to build electronic interfaces between living cells and mechanical devices, which will revolutionize medicine, biology and electronics. The border between biological life and artificial life will soon blur and may even eventually disappear.

QUANTUM EVOLUTION AND THE CELL

But quantum evolution is far older than the experimental demonstration of adaptive mutations. In the last chapter, I proposed that quantum evolution was the key to life's origins. But its role did not stop there. Quantum evolution has probably played an essential role in the development of life on Earth throughout its three and a half billion-year history.

Darwinian natural selection is generally considered the sole process that has guided evolution through the multiverse of biological design towards viable and successful creatures. You will remember that the key to natural selection is the action of the Grim Reaper who separates the wheat from the chaff of natural variation. The development of complex organs, like the eye, may seem incredible but, so long as every step along their evolutionary path provided a distinct advantage, each could have been captured by natural selection. The evidence for these evolutionary paths is that their stepping stones can still be identified, both in fossils and even in living animals. The successful designs, evolved millions of years ago, continue to provide the same advantages for living creatures today.

Nevertheless, problem areas remain, particularly when we come to examine molecular evolution. As described in Chapter Four, one problem is the relationship between major protein families. Although evolution within protein families (such as the globin family) can be traced through a number of antecedent proteins present in living creatures, finding the links between these galaxies of protein families is much harder. Molecular evolution appears to have proceeded though a series of small steps, but also seems to have taken big leaps. These leaps are very difficult to account for in terms of the gradual change from one species to another envisaged by Darwinian evolution. Without natural selection to guide evolution, we are back to the problem of monkeys and typewriters. The probability of big random changes generating anything but nonsense is extremely remote.

Another problem described in Chapter Four is the existence of biochemical pathways that appear irreducibly complex. The cell makes the energy storage molecule ATP (adenosine triphosphate) by adding a phosphate group onto ADP (adenosine diphosphate). ADP is similarly synthesized by adding a phosphate group onto AMP (adenosine monophosphate). However, the biosynthesis of AMP from its precursor, ribose-5-phosphate, involves thirteen independent steps involving thirteen different enzymes (which we will represent as: $A \rightarrow B \rightarrow C \rightarrow D \rightarrow E \rightarrow F \rightarrow G \rightarrow H \rightarrow I \rightarrow J \rightarrow K \rightarrow L \rightarrow M$, where A is ribose 5-phosphate and M is AMP). Each enzyme involved in this pathway is absolutely essential for the biosynthesis of AMP. Darwinian evolution requires this complex system to have evolved from something simpler but, as far as we know, nothing simpler works.

In *Darwin's Black Box*[8] Michael J Bethe proposed that these metabolic pathways could have been put in place only by God. However, I do not believe we need resort to pre-Galilean science to explain their existence. Quantum evolution may instead provide the answer and also to the related problem of the apparent big hops in molecular evolution. Consider again the gene encoding the functional beta-galactosidase enzyme. As we discovered, the encoded enzyme's ability to transform lactose entangles the gene with the environment of the cell. For fully functional genes, this entanglement will cause the cell to perform a dense series of quantum measurements on its DNA. But these continuous measurements tend also to invoke the quantum Zeno effect, which freezes the dynamics of quantum systems. Continuous quantum measurement of active genes will therefore tend to freeze their DNA dynamics and promote evolutionary stasis. In effect, active genes are continually being nailed to their functional sequence by quantum measurement.

But now consider what might have happened if, in the distant past, the beta-galactosidase gene (or any other functional gene) was accidentally duplicated during its replication.[9] The duplicate gene would have no longer been essential because the original copy was sufficient to meet the cell's needs. Still subject to the replication errors that generate mutations, the copy would have been without the pull of natural selection, continually dragging the sequence back to its original (functional) state – thus its DNA sequence would have been allowed to *drift*. Eventually, the duplicated gene would have suffered a hit from a mutation that inactivated its original function and broke the chains of entanglement which allowed continuous quantum measurement. The quantum Zeno effect would no longer be able to nail the sequence to a classical reality, allowing the gene to descend into the quantum world. Thereafter, the gene would have continued its drift but within the quantum multiverse (if you will allow me to return to the Many Worlds interpretation of quantum mechanics either as representing reality or, at least, a convenient metaphor for something deeper) of gene sequences, rather than the classical world. Each subsequent mutation would have taken place – not as a single mutation, but as a quantum superposition of all possible mutations. The superposition would have grown, branching like a tree, into the multiverse of all the trillions of possible DNA sequences that could lead off from the original gene.

In fact, a significant proportion of the genomes of all organisms may, like our hypothetical duplicated sequence, be capable of this kind of quantum sequence drift. Up to about ten per cent of most bacterial genomes is thought to represent *junk DNA*, sequences that have become non-functional for one reason or another. The proportion of junk DNA is far higher in more complex genomes such as our own. Perhaps as much as ninety per cent of our own DNA may be junk sequences. This will be invisible to the cell's quantum-measuring devices, allowing their mutational events to drift unnoticed into the quantum realm.[10]

The only event capable of halting our duplicated gene's drift through the quantum multiverse is the establishment of another chain of entanglement with the cell's environment. This would have happened whenever the superposition included within it a sequence that encoded a novel enzyme capable of interacting with the cell's environment, such as an enzyme able to utilize a new substrate. At that point, a new chain of entanglement would have been formed to create a new channel for quantum measurement. The gene superposition would have collapsed into a single classical sequence once again.

The gene sequence that became entangled with its environment might have encoded a novel enzyme, ancestor of a new gene family. That enzyme might have conferred a unique metabolic faculty on the cell, one vital for the subsequent evolution of living creatures. But, although the vital sequence *performed* the essential measurement that led to (wave function) collapse, it would have had no greater claim on the subsequent reality than any of the trillions of useless sequences also present within the superposition. If this was all that was going on, the probability for the novel sequence's emergence would be no greater than if given a classical evolutionary scenario.

At this stage, we could invoke the anthropic multiverse (Many Worlds interpretation) to get us out of this difficulty. All we need to propose is that our existence is contingent upon the evolutionary innovation conferred by the novel sequence. We could similarly account for the emergence of all the major families of protein sequences. Our presence here today guarantees that we live in the lucky universe that hit the jackpot, scooping all the enzyme families necessary for intelligent life's subsequent emergence. Although this would account for the facts, particularly the absence of viable intermediate sequences, I find the explanation just as unsatisfying now as in the last chapter.

An alternative explanation would again invoke the power of the quantum Zeno effect (and its inverse) to enhance quantum probabilities. Junk DNA that accumulates mutations may evolve in the quantum realm, constituting a kind of quantum forest of possible gene sequences. As above, any one may encounter an environmental entanglement that generates measurement – a proto-enzyme activity – but this proto-enzyme would (as in the origin of life scenario) emerge unscathed from measurement. However, occasionally a chain of proto-enzymes acting in concert may confer a new metabolic capability on the cell, amplifying the quantum state irreversibly to the classical level.

Such a chain of measurements could have led to the biochemical pathway $A \rightarrow B \rightarrow C \rightarrow D \rightarrow E \rightarrow F \rightarrow G \rightarrow H \rightarrow I \rightarrow J \rightarrow K \rightarrow L \rightarrow M$ to make AMP. Each proto-enzyme along the route would have performed a quantum measurement but, in isolation, the enzyme emerging from the measurement would have continued to evolve in the quantum multiverse. However, once a chain of proto-enzymes had come together to confer a novel metabolic activity on the cell, it would be the end of this line of quantum measurement. The enzyme chain's quantum state would irreversibly collapse into a classical state. We then have the (by now familiar) chain of measurements leading to a classical state, and the power of the inverse quantum Zeno effect to increase the probability of reaching that state. Each enzyme along the path could have performed the same kind of quantum measurement role performed by the Polaroid lenses in our experiment in Chapter Eight. Just as a series of lenses inserted into a light path can enhance photon transmission probability, so a series of enzymes inserted into a metabolic path may have enhanced the probability for evolution of the entire pathway.

Once the metabolic pathway emerged from the superposition, Darwinian natural selection would have guided its subsequent improvement. These subsequent Darwinian changes would have left their record in the molecular clock as the gradual modification of members of gene families. But the initiating event – the emergence of the entire pathway and its set of enzymes – would have left no trace (because it took place in the quantum multiverse). In this way, both the existence of diverse gene families, and complex metabolic pathways become comprehensible in the terms of quantum evolution.

It is important to realize that quantum evolution does not replace natural selection. Instead, at critical evolutionary junctures, it shifts natural selection into the quantum realm. Darwinian evolution then takes place in the quantum multiverse. In this way, even lowly E. coli cells may have a certain control of their destiny; a control denied to inanimate objects. This is what makes living organisms special. They are able to use quantum measurement to perform *directed actions* and one of those actions is quantum evolution.

QUANTUM EVOLUTION TODAY

The early stages of precellular and cellular evolution were, I believe, dominated by quantum evolution. Quantum evolution drove the living cell towards increasing environmental entanglement and thereby greater complexity. Eventually, a certain size and complexity were reached when no further increase was possible without loss of quantum coherence, and thus quantum evolution. This impasse lasted for billions of years before the next step, the development of multicellular life, arrived six hundred million years ago. Living creatures were then constructed from colonies of unit cells. Communication between cells allowed the co-ordinated behaviour of the entire colony, and so animals and plants evolved. Genetic changes in germ-line tissue were now unlikely to penetrate directly to the organism's environment. The entanglement between the genetic material and the cellular environment – necessary for quantum evolution – was now curtailed. Nevertheless, quantum evolution may still play a role in microbial evolution, and also within multicellular creatures when a single cell's activity becomes dominant.

Professor Hall has proposed that adaptive mutations are involved in the acquisition of drug resistance by harmful bacteria. Drug resistance is an enormous problem in treating many infectious diseases and particularly in the biggest killer of them all – tuberculosis. Tuberculosis or TB, kills more people than any other pathogen – about three million each year and rising. Most deaths occur in the developing world where access to drug treatment and disease control is still largely the prerogative of the wealthy. In the West, the disease has been curable since 1944 when it was discovered that the tubercle bacillus was susceptible to the drug streptomycin. However, early successes were followed by relapses with drug-resistant strains. The tubercle bacillus mutates to become streptomycin resistant only very rarely: about one resistant mutant will be found in 10^8 (100 million) sensitive bacilli. However, as patients may harbour up to 10^9 bacteria in their body during active disease, it is hardly surprising that treatment with the single drug will inevitably select those very few mutants resistant to streptomycin.

A solution to the problem of drug-resistance was soon discovered: multidrug therapy. Patients were treated with four drugs including two key antibiotics: *rifampicin* and *isoniazid*. Normally, mutations to rifampicin resistance occur in only about one cell per 10^{10} sensitive cells; mutations to isoniazid resistance occur in about one in 10^8 cells. Resistance to the other drugs used in TB therapy occurs more frequently, but for a strain to overcome treatment with all four, it must acquire four independent mutations. The combined mutation rate should be a product of the individual rates, giving an expected frequency for acquisition of resistance to all four drugs, of less than one in 10^{24}. To have a reasonable chance of selecting only a single four-drug-resistant mutant, you would need a population of sensitive TB bugs in excess of one trillion trillion (10^{24}). This would weigh more than one billion kilograms. Thankfully, there is nothing approaching that number of TB bacilli in the entire world. So, provided patients are being treated correctly, resistance to all four drugs should be prevented. Yet multidrug-resistant strains have now emerged in many parts of the world. One of the scariest strains, *strain W*, emerged in New York and has caused more than one hundred cases of TB with a mortality greater than seventy per cent (even under treatment). This strain is resistant to seven of the front-line drugs used for TB treatment and, remarkably, each resistance is caused by independent gene mutations. Strains like W are in danger of turning back the clock fifty years to the days when TB was incurable.

The phenomenon of multidrug resistance in TB is a major problem for its control throughout the world, but I believe it also poses a problem for neoDarwinian evolution. The acquisition of multidrug resistance for strain W must have involved a series of seven mutations: sensitive strain → resistance to one drug → resistance to two drugs →→→→→ strain W, resistant to all seven. Darwinian natural selection could have guided the evolution of the strain W through this series of mutations, *but only if each step along the path provided a selective advantage to the tubercle bacillus*. However, if a patient is undergoing multidrug therapy (as all TB patients in the US are meant to), single mutations are of no value to the bacillus; the other drugs will still kill the pathogen, but only so long as the patient takes all his tablets. This is a very important since 'treatment compliance' is a big problem in TB control. The full course of TB treatment lasts six months, but patients usually feel much better after the first few weeks. After that the patient may forget or not bother to complete the course of treatment. If he or she continues with just one drug, then the resulting monotherapy might select a single-resistance mutant strain.

Although successive transmission of the strain from one poorly compliant patient to another could account for the eventual evolution of strain W, I do not believe it very likely – even given the deficiencies of the American Public Health Program in the 1970s. For a start, this should have generated a whole series of intermediate resistant strains that would also have turned up in American TB laboratories. A TB strain that looks the same as strain W on genetic fingerprinting is actually quite common in the United States (and throughout the world), but it is sensitive to all seven drugs. A strain that resembles strain W but is resistant to three of the seven drugs has also been isolated. But most of the other presumed ancestors of strain W are conspicuous by their absence. The evolution of the sensitive version of strain W to its highly resistant descendant may have involved a giant leap rather than small Darwinian steps.

A possible explanation for strain W's rapid evolution is adaptive mutations and quantum evolution. Hall (and others) have shown that adaptive mutations tend to emerge in cells unable to grow for long periods: when remaining viable but dormant. It is in this state of dormancy that quantum coherence may be tightest. And the TB bacillus is the champion of bacterial hibernators. About one third of the world's population is TB-infected but, in most people, it persists in the lungs in a non-growing or dormant state where it does no harm. Work by a PhD student (now Dr Kiran Ghanekar), in my laboratory at Surrey University, has uncovered an increased rate of mutations to drug-resistance in a relative of the tubercle bacillus, whilst persisting in this non-growing state. We do not yet know whether these mutations are truly adaptive (that they occur only under selective conditions – as unfortunately the genetic tools to test this are still somewhat primitive for this group of bacteria) but they were only detected in cells incubated for several weeks in non-growing conditions.[11] They do, at least, have one of the key features of adaptive mutations.

Adaptive mutations (and, so I believe, quantum evolution) may be responsible for making TB – already the biggest killer in the world – even more dangerous. But adaptive mutations may also be taking place within our own body, where they may be involved in that most feared of western diseases: cancer.

Cancer is caused by cell mutations which allow them to grow uncontrolled until they form the destructive mass of cells that we call a tumour. Yet it is not easy to persuade our cells to divide endlessly. All our cells are pre-programmed to die after a certain number of generations, usually about thirty. Indeed, most of our cells are pre-programmed to kill themselves by a cell suicide process (apoptosis) unless constantly reminded not to do so. For a cell to overcome these inherent checks on growth and thus form a tumour, they must acquire at least two, often three or more, mutations. The chances of two or three mutations occurring in the same cell (like mutations to multidrug resistance in TB) should be infinitesimal. People with inherited susceptibility to cancer are unfortunate in that they have inherited one of these mutations (such as the breast cancer susceptibility gene, *BRAC1*) directly from their parents. Their cells need only acquire a single extra mutation to escape growth control and generate a tumour. However, many cancers occur without evidence of inherited susceptibility, yet still require the acquisition of two or more mutations within a single cell. Professor Hall has proposed that adaptive mutations may be the source of these. If, as I am proposing, quantum evolution is the source of adaptive mutations, then cancer may be the phenomenon's downside.

So, in addition to its role in promoting life's emergence and the evolution of the earliest cellular life, quantum evolution may be alive and well in modern organisms. It is perhaps unfortunate that despite its honourable history, today's manifestations of quantum evolution appear, from a human perspective, mostly negative. However, we have not yet discussed one final, and I believe critical, role the quantum cell plays in our world today. Quantum mechanics may be the source of that most prized possession of humankind: the conscious mind.

13

Mind and Matter

On 18th July 1897, the *Seattle Daily Times* ran the story, 'At 3 o'clock this morning, the steamer *Portland* from St Michael for Seattle, passed up the sound with more than a ton of solid gold on board . . .'. The news flashed around the globe and the greatest gold rush the world has ever seen headed for the Klondike.

The late 1890s had seen one of the deepest global depressions of modern times. Millions were laid off; thousands of families were evicted from their lands, and the homeless left to starve in the streets. And then the SS *Portland* steamed into Seattle harbour with its cargo of bright gold. Tales of snow-covered fields sprinkled with gold dust swept across the world and, within days, tens of thousands of men and women sold what possessions they had to book passage to the Klondike.

Most were not professional prospectors, but unemployed bank clerks, farm labourers, dentists: anyone young and desperate enough to chance their luck. Few knew anything about gold-prospecting or that the journey to the Klondike was among the most arduous in the world. Many headed north to the south-east Alaskan town of Dyea and the start of the thirty-two-mile-long Chilkoot Trail, their first and harshest test. Prospectors had to carry a year's supply of food for the journey which, together with their equipment, weighed about a ton. The first stage was a 3,550 foot climb up the mountainside, each prospector having to make as many as twenty successive trips to haul all their load. And that was only the beginning. Before they reached the Klondike, they had to travel for months across snow-capped mountains, frozen lakes and glaciers, and endure temperatures that dropped to −50°C. Many became so exhausted that they sold or abandoned their goods and turned back. Many others

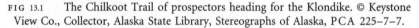

FIG 13.1   The Chilkoot Trail of prospectors heading for the Klondike. © Keystone
View Co., Collector, Alaska State Library, Stereographs of Alaska, PCA 225–7–7.

died on the trail; falling into crevasses, buried under avalanches or mur-
dered by bandits. Those that made it founded the town of Dawson which
still stands today, on the banks of the Klondike. Though both sober and
respectable today, in the 1890s it was notorious – the most northerly
outpost of the Wild West lifestyle where some of the surviving prospectors
lost their remaining money and possessions to the town's thieves, gam-
blers and con-men.

The above photograph, taken by an anonymous photographer, is a
most striking depiction of the power of the human will. The whole history

of man's struggle to impose himself on a hostile environment seems written in that thin black line trudging over the Chilkoot Pass. Our will – our ability to make decisions and direct our own actions – has surely been our most valuable (and dangerous) asset in that long road from the primeval forests to our modern cities. Without it, we would never have fashioned tools, planted crops, tended herds, built cities or forged weapons to destroy crops, herds, cities and people. There would be no civilization, no buildings, no paintings, no music – and, of course, no books about the origin of all these. Each of these achievements takes an enormous effort against the tide of inevitability. Our will is surely the most striking manifestation of life's ability to perform directed actions. But where does it come from?

Consider the scene in the picture as it might have been witnessed by the imaginary alien spacecraft introduced in Chapter One. At daybreak, it would spot a thousand tons of amorphous material – perhaps a mass of 'rock' (though in reality people and supplies) – lying at the foot of a mountainside. By dusk, that same material would have been elevated by several thousand feet. The spacecraft would have been left with a problem: to explain how this mass of *rock* managed to increase its potential energy so enormously as to elevate itself up the mountainside. It would have first looked for some external agency acting upon the rocks, capable of raising them several thousand feet against the force of gravity; it would have found none. It would next have attempted to account for the feat in terms of the internal dynamics of the system, perhaps as some spontaneous physiochemical reaction. As we discovered in Chapter Five, Newtonian mechanics and its statistical cousin, thermodynamics, govern the motion of inanimate material. To account for the Chilkoot climb in purely mechanical or thermodynamic terms, the alien spacecraft would have had to suppose that all the molecules in the rocks and their surroundings were so arranged that their random bumping and jostling (which is, of course, all there is to thermodynamics) caused the entire rocky mass to ascend spontaneously up the mountainside. Is such a view tenable? Could random mechanical and thermodynamic forces have accounted for the climb over the Chilkoot Pass?

I am sure that you will not be surprised that I believe the answer to that question is no. The alien spacecraft would recognize, in the Chilkoot scene, the same signature of life that it spotted in the bird soaring into the sky or a salmon leaping a waterfall: living organisms' ability to initiate directed actions. But how does the human will cause the motion of bodies on such massive scales? To attempt to answer this, we need to explore how we *will* our bodies into action.

HOW NERVES MOVE MUSCLE

In Chapter Five, I described how our voluntary muscle cells contract when we kick a football. The same muscle contraction similarly accounts for the ability of the Klondike prospectors to drag themselves and their supplies up a mountainside. But what causes muscles to contract in response to will? How do we *will* matter (our muscles) to move?

We have already seen (Chapter Five, Figure 5.1) how the mechanical energy for muscle contraction is provided by ATP's hydrolysis by myosin molecules. But what causes myosin to hydrolyse ATP and thereby initiate muscle contraction? The immediate answer is calcium. Raised calcium levels trigger the enzymatic activity of myosin. The raised calcium levels are caused by a release of calcium from intracellular calcium stores in response to *electrical depolarization* of the muscle cell-membrane.

Most cell membranes are electrically polarized, with more positive ions outside the cell than inside, leading to a negative voltage across the cell-membrane. However, membranes can *depolarize* if positive ions are allowed to travel through membrane pores to neutralize this voltage difference. Muscle and nerve cells have special pores (voltage-gated ion channels) that open and close in response to changes in the voltage difference across cell membranes. They remain closed as long as the voltage difference is sufficiently negative, but pop open whenever the electronegativity drops below a critical threshold. Yet throwing the channels open only lets in more positive ions to cause a further drop in the membrane voltage, thus popping open more voltage-gated channels and precipitating further rapid cascade of depolarization. This accelerating membrane depolarization (action potential) stimulates the release of intracellular calcium stores within muscle cells and so initiates muscle contraction.

The next backward step in our chain of causation from the prospector's leg, is to understand what causes the initial electrical depolarization opening the voltage-gated channels in his leg muscle. This is where nerves enter the picture. Motor nerves (nerves that stimulate muscles) terminate in structures called synaptic knobs which abut against muscle cells at neuromuscular junctions. The synaptic knob releases a chemical signal (neurotransmitter) into the fluid-filled space between the nerve-cell ending and the muscle cell: the *synaptic cleft*. Different types of nerve endings release a varied bunch of neurotransmitter signals, but most motor nerves release a neurotransmitter called acetylcholine. Muscle cells have acetylcholine *receptors* embedded in their membranes that act as *ligand*[1]-gated ion channels. Whenever these receptors capture an acetylcholine molecule (released by the nerve cell's synaptic knob), they briefly open a channel for sodium ions to flow into the muscle cell. If enough ligand-gated channels are opened to allow in many sodium ions, then the membrane potential is reduced below the threshold needed to pop open the voltage-gated ion channels and will initiate the action potential.

So the voltage-gated ion channels that kick the muscle into action, are opened up by the action of another set of channels – the ligand-gated ion channels – which respond to neurotransmitters released into the synaptic cleft. Our next link is to understand what makes the motor nerve cell release those neurotransmitter molecules. The motor nerve cell's synaptic knob is full of tiny vesicles filled with thousands of acetylcholine molecules. The nerve cell discharges the contents of these vesicles into the synaptic cleft, whenever an action potential arrives at the synaptic knob.[2]

Action potentials are fundamental to nerve action, so we need to take a closer look. Nerves (or neurones) are very long cells (they can be more than a metre long) with a spidery cell-body at the cell's head-end, connected by a long axon to its tail-end: the nerve ending or synaptic knob that releases neurotransmitter molecules.[3] Signals are communicated along nerves by action potentials that travel along the axon from the cell-body to the synaptic knob. The axon resembles a thin wire, so you might think that nerve impulses would be transmitted by a flow of electrons, just as electrical signals are passed down a metal wire. But you would be wrong. Nerve transmission is very different. Like muscle cells, resting neurones have a voltage difference across their cell-membrane, maintained by a sodium pump pushing positively-charged sodium ions out of the cell. Normally the voltage difference is about −65 millivolts (positive outside, negative inside), which may not sound like much but since cell-membranes are less than a thousandth of a millimetre thick, in fact amounts to a staggering 13,000 volts per centimetre. In another similarity to muscle cells, neuronal membranes have voltage-gated sodium channels that open up whenever the voltage drops below about −40 millivolts. To see how action potentials work, imagine first that a few voltage-gated channels are already opened at the cell-body (the head) of the nerve. Positively charged sodium ions will rush in through the channels to reverse the voltage difference across the membrane, causing membrane depolarization. When the voltage dips below −40 millivolts (for this to happen thousands of channels must open) then adjacent voltage-gated ion channels are provoked into popping open. This will cause another surge of sodium ions to enter the cell and the further depolarization will stimulate the next set of membrane channels along the axon to open their doors. This process will continue as a wave of membrane depolarization – the action potential or nerve impulse – that travels from the cell-body along the nerve axon, at a rate of about 100 metres per second, until it reaches the synaptic knob.

But we have so far just imagined the initial membrane depolarization caused by the opening of a few sodium channels. What normally opens these channels, causing the neurone to *fire*? Mostly it is other nerves. The cell-body of a motor nerve is located in the spinal cord. It possesses long spidery extensions called dendrites which are targets for synaptic knobs of connecting nerve cells. The *upstream* synaptic knobs release their neurotransmitter load into the synaptic cleft, to be picked up by receptors on the dendrite extensions of the motor nerve-cell body. How the motor nerve-cell body interprets the neurotransmitter signal varies greatly, depending upon the type of neurotransmitter. Some will open ligand-gated ion channels, while others close them. If the cell receives enough 'open' signals, sufficient ions will enter the cell-body to decrease its membrane potential below the critical threshold of about −40 millivolts and pop open the voltage-gated ion channels to initiate the action potential.

So the neurone is a democrat. It will decide whether or not to fire on the basis of the balance of neurotransmitter votes it receives.

HOW OUR BRAIN MOVES NERVES

Each nerve cell is an information-processing centre: it has an input (usually synaptic signals from other nerves), an information-processing centre (the cell-body) and an output (to release neurotransmitter or not into the synaptic cleft). The cell-bodies of most voluntary motor nerves (whose nerve-endings terminate at neuromuscular junctions) are located in the spinal cord, where they form synapses with sensory neurones (mostly through an intermediary interneurone) and neurones from the brain. If unlucky enough to stand on a nail, your leg muscles will immediately contract to withdraw your foot, in an action known as the *flexor reflex*. This reflex is initiated by a signal from sensory nerves in your foot, which registers the breaking of the skin (sensory nerves have modified cell-bodies that directly register physical signals, such as light, heat or touch, instead of receiving signals from another cell). The nerve-signal races up your leg and enters your spinal column, to be transmitted (via an excitory interneurone) to the motor neurone. The signal is thereby transmitted from your foot to your calf and thigh muscles, without your brain's involvement (though a pain signal is sent to your brain, it arrives after the reflex initiation).

However, our prospector's first step up the mountainside was unlikely to have been a reflex (unless, of course, he happened to tread on a nail). It was voluntary, and voluntary actions are initiated in the brain. Undoubtedly the most complex biological system that has ever evolved on this planet, the human brain may indeed be the most complex organized system in the entire universe. However, the observant reader will surely have spotted that the remaining pages in this book are few, and that the problems of the human brain are many. They will be sceptical of any attempt to tackle that bastion of anatomy, neurophysiology, psychology and indeed philosophical speculation, in the remainder of this book. They are right to be so. The brain and its most mysterious occupant – our own consciousness – is a vast topic for an entire volume, let alone a single chapter. There are many excellent and interesting texts (some recommended in the bibliography) which deal with the brain and its functions in the detail more appropriate to that topic's complexity. So I will limit our exploration of the brain to the very minimum needed to explain why I think we need quantum mechanics to account for the actions of our gold prospector.

Our brain consists of about one hundred billion (10^{11}) neurones and about a trillion (10^{12}) non-nerve cells, known collectively as *glia*. A huge amount of evidence has accumulated to indicate that it is within the thin sheet of cells of the brain's cortex (the grey wrinkled layer that forms the brain's outermost surface – only about six cells thick) that most information processing takes place. The Canadian neurosurgeon Wilder Penfield, from the 1930s to the 1950s, performed pioneering studies, mapping those regions of the cortex involved in various sensory and motor activities. Penfield was able to electrically stimulate discrete areas of the cortex of patients undergoing brain surgery. Remarkably, because the brain has no pain receptors, the operations could be carried out under local anaesthetic, allowing the patients to describe the sensations experienced when particular regions of the cortex were stimulated. Penfield was able to map the cortical areas involved with touch-sensory perception (when these areas of the *somatosensory cortex* were stimulated, the patients would experience a tingling sensation); visual perception (when the *visual cortex* was stimulated, the patients would *see* bright lights); and voluntary movement (when the *motor* cortex was stimulated, the patient's arm or leg would twitch). Even more remarkable was Penfield's finding that when he stimulated the brain area, known as the temporal lobe (located on the brain's lower surface, under the temporal bone), patients would hallucinate or recall long forgotten incidents (the patients would say something like, 'I feel as though I were in the bathroom at school'). It appeared that Penfield was reactivating long-forgotten memories stored in the temporal lobe.

So the nerve impulse that started our prospector up the mountainside would have had its origin in a neurone or assembly of neurones within his motor cortex. But what caused the critical neurone to fire? Like most other nerve cells, it must have had many inputs from other neurones. The dendrite extensions of brain nerve cells are massively branched, forming synapses with thousands of other nerve endings to form a densely integrated neurone network. The critical motor neurone in our prospector's brain would have certainly had many inputs from neurones in the somatosensory cortex, the visual cortex and many other brain regions.

Each input would have had its own input nerve cell. That nerve cell, in turn, must have its input neurone, and that neuron must have its input neurone, and so on backwards through an infinite regression of outputs and inputs. Where does the buck stop? Which neurone takes the decision whether or not to fire our gold prospector up the mountain?

If man were a machine, a robot, it would be easy to envisage a simple stimulus-response mode of muscle firing. Man sees mountain, eye sends signal to brain (along sensory nerve), brain sends signal to muscle. Voluntary movement is clearly not this kind of reflex action, but must depend on more complex signalling and information-processing between the stimulus and response. Consider the gold prospector, paused at the foot of the Chilkoot Pass. With a hundredweight of supplies on his back, he looks up at the expanse of snow and ice and thinks of the long weeks of cold and hardship he must endure before reaching the Klondike. He sees visions of gold in the snow, but also sees smoke from the log fires burning in the cabins below. Does he take his first great stride up the slope or instead turn back towards warmth, comfort and failure? If we could have asked one prospector who made it to the top why he took that first step, he would have told us of his dreams: the smiling faces of his wife and children when he returns laden with riches; the house he would buy, the clothes he would wear, his proud expression as he strides triumphantly down the main street of his home town. He would certainly not describe his actions in the same terms as he would explain how he leaped up after treading on a nail. He would have assured us that he had made a *conscious* decision to climb that hillside. Was he right?

HOW CONSCIOUSNESS *MOVES* OUR BRAIN

We feel that we consciously *will* our voluntary actions, but how can something as intangible as consciousness move our muscles? To initiate muscle contraction, our conscious will must stimulate neuronal firing within our brain's motor cortex. But the opening of ion channels in the cell-membrane of nerve cells causes neuronal firing. These ion channels are made of the same protein one finds in a peanut. The power of our will (without the aid of muscles) cannot move a peanut's proteinaceous matter, so how can it move the same kind of matter (ion channels made out of protein) when it is inside our brain? How does mind move matter?

This question, often referred to as the mind-body problem, goes back at least as far as the Greek philosophers. You may remember from Chapter One that Aristotle considered the body to be made of matter but the immaterial *psyche* (soul) initiated the movement of that matter. Aristotle believed that the heart was the seat of the soul, the brain's function merely to cool the blood. That most influential Roman physician, Galen, taught the modern view that the brain was seat of knowledge, intelligence and will. Galen proposed that voluntary movement is initiated by the motion of *humours* within the brain's fluid-filled ventricles and that these disturbances travel down the nerves – which he thought to be hollow fibres – to the muscles.

The concept of the brain as a mechanical pump appealed to the mechanists of the seventeenth and eighteenth centuries, but even their champion, Descartes, could not accept that the basis of all human actions was mechanical. Instead he advocated what has come to be known as the *dualist* tradition, that the human mind is composed of the material brain, but also an immaterial mind or soul, whose spiritual substance lies outside science. The brain's job was to perform all the mechanical tasks we share with beasts, like walking or eating, but our incorporeal mind was held to be the source of thoughts, feelings and conscious actions.

Most modern scientists reject dualism, instead embracing *monism*: that the stuff of mind is the same as that of brain, matter. Many consider that the brain works in essentially the same manner as any computer, but that it is more complex and wired somewhat differently. Let us examine that hypothesis next.

THE COMPUTING BRAIN

Our brain certainly has impressive computational skills but is it a computer? To answer this question, we need to know just a little about how computers work. All modern computers are composed of tiny electrical circuits (called bits = Binary digIT) which send a signal, either ON or OFF. A logic gate is a circuit that combines these bits to perform a particular logical operation. For example, an AND gate has two input signals and a single output. If both its inputs are ON, it switches its output to ON. An OR gate will switch ON if either of its input circuits are ON. Computers perform calculations by combining the logical operations performed by gates to perform the necessary additions, subtractions, multiplications, etc. and arrive at an answer. The sequence of logical operations used to perform a particular calculation are called *algorithms*.

A major difference between the neurones in our brain and modern computers is that a computer logic gate has few input circuits (usually two), whereas a single neurone may receive input from thousands of upstream neurones. Yet it may still perform the same kind of algorithmic computation as a computer: firing only if all input neurones are ON (an AND gate), or firing if any input is ON (an OR gate). A complex network of gates may perform the detailed calculations necessary to decide whether or not to initiate a certain action. The *decision-making neurone*[4] of our gold prospector will have received signals from his visual and somatosensory cortices containing information concerning the steepness of the slope, the weather, temperature and the tiredness of his limbs. These inputs will have been processed during their passage through a complex network of neuronal gates before arriving at an answer: to either stimulate or suppress the critical neurone's firing. Returning to the Chilkoot, let us imagine that the weather was particularly fierce on the morning of our prospector's decision, so most of the sensory inputs' synaptic knobs released inhibitory neurotransmitters towards the decisive motor nerve. Stimulatory signals may have arrived from the brain's other regions, perhaps those concerned with memory. Our prospector's temporal lobes may have held images of a hungry child or a wife dressed in rags and these would have been processed to send signals to urge him forward in his search for gold. Once the decisive neurone had received all its inputs, it may have performed a simple calculation: add up all the stimulatory signals, subtract the inhibitory and if the answer generates a membrane potential less than −40 millivolts, get up that mountainside.

But what then is the purpose of the prospector's consciousness, if all his decisions are determined by brute neuronal calculations? Why does he need to be *aware* of his actions if their cause is neuronal number-crunching? Wouldn't an unconscious computer do the job just as well? A fundamental principle of computing is that the algorithms performed by one algorithmic computer can in principal be run on any other algorithmic computer.[5] If a computer were built to go through the same algorithmic routine as those utilized by our prospector's brain, would it also be conscious? Many computer scientists take the view that it would. They consider that consciousness is simply a by-product of extremely complex computation; and that any computer which could perform the algorithms a gold prospector performs, would inevitably become conscious. But why should it? It would serve no function. Consciousness would have no role to play in the computer's decision-making process. A conscious computer would perform the same calculations and make the same decisions as an unconscious. Does consciousness similarly play no active role in the decision-making process taking place within our brain? Would an unconscious zombie make the same decisions as our gold prospector?[6] Are we just automatons that happen to be *aware* of our actions because of some evolutionary accident? In the words of the evolutionary biologist T. H. Huxley, is consciousness like the 'steam whistle which accompanies the work of a locomotive [but] without influence upon its machinery'?

INITIATING ACTIONS

Most readers would, I guess, like myself, be reluctant to deny the role of our conscious free will. We all feel strongly that there is a mind inside our head with the power of volition over our actions. Yet experiments performed by the American neurobiologist, Benjamin Libet, of California University, profoundly challenge this belief. With the neurosurgeon Bertram Feinstein, Libet carried out a series of fascinating studies on the timing of sensory perception and motor actions. Although many of his experiments were performed *intracranially* on patients undergoing brain surgery, it was a simpler, less invasive procedure that yielded his most startling finding. Libet asked normal healthy subjects to flex their finger at a time of their own choosing. He placed electrodes on their scalp, to record the brain's electrical activity associated with this action. The subjects would also record when they thought they had initiated the action by noting the position of a rapidly rotating clock-hand. Libet would monitor the motor action by recorders attached to the person's limbs. It takes only a few milliseconds for a nerve impulse to pass from brain to muscle, so this is a pretty good marker for the initiation of motor neurone firing in the brain.

Subjects reported their awareness of *making a conscious decision to move*, roughly two hundred milliseconds before the action was recorded at their muscle. The timings indicate that there is generally a delay of two hundred milliseconds between the time we become aware of our intention to perform a conscious action and firing up the appropriate motor nerve. However, what was much more surprising was that Libet routinely detected neuronal activity in the brain associated with the voluntary action, a full three to four hundred milliseconds (nearly half a second), before the time the subject reported he or she had made the decision to move. These apparently voluntary actions were initiated well before the subjects knew they had made any conscious decision to act.

At first sight, this experiment seems to demonstrate that we are automatons. Voluntary actions are really unconscious acts which we retrospectively become aware of. In this view, the decision to send our prospector up the mountainside was made well before he knew where he was going. His brain performed a complex calculation of the pros and cons of each option, sent him up the mountainside (or not), and only later made him aware of his actions.

Is free will therefore an illusion? Are we slaves to our brain's unconscious neuronal activity? Where does that leave our sense of responsibility and conscience, our pangs of guilt or pride in our actions; are they all delusions? Have we the right to punish wrongdoers, if they could not help their actions? But following that argument, we are unable to help our own actions in punishing them.

Man, as an aware but helpless robot, is a depressingly bleak vision of the human condition. Fortunately, it is unnecessary. Libet's experiments did not compel *him* to abandon the notion of free will. He proposed instead that 'consciousness acts to modify or veto actions that are initiated unconsciously.' Neuronal activity may precede a conscious decision to act, by three to four hundred milliseconds; but (and crucially) there was still a gap of two hundred milliseconds between the awareness of a conscious intention to act and the initiation of the motor impulse. Libet proposed that it is in this motor lag period that consciousness can have an influence on voluntary action. Voluntary actions may be initiated unconsciously but, before they are consummated, consciousness can intervene to veto or reinforce the action. Libet found evidence for this by observing that the kind of neuronal activity which is often followed by motor action, was sometimes aborted before that action was completed.

This explanation makes a lot of sense in the terms of my own experiences. I can remember watching that particularly startling scene in *Alien*, when the monster bursts out of John Hurt's stomach. Like many others in the audience, I 'started' to scream, only for that action to be vetoed by my conscious mind (which *knew* I was in a crowded cinema). I am sure there are many similar occasions when your conscious mind similarly asserted itself, to interrupt a potentially embarrassing voluntary action. We often describe those who are less able at this skill as people who are always '*putting their foot in it*'.

So much of the computational work concerned with initiating the motor actions that took our prospector up or down the mountainside would have been initiated unconsciously, but there was still a window (of about two hundred milliseconds) when his conscious mind could have intervened to choose or reject any action. That brief *window of consciousness* is the entry point for our free will. Let us next explore what can be seen through that window.

BINDING OUR THOUGHTS

If we were to ask our prospector why his decision to climb the mountain or not needed to be conscious, he would have no difficulty answering. He would have described all the factors that could influence his decision: the snow on the mountains, the wind's howling, the cold, the weight of his backpack, the tiredness of his limbs, his likelihood of success – all the dangers and the potential rewards. Whilst making his decision, his conscious mind would have been aware of all these varied inputs as a continuous stream of information. How did all that data fit into his conscious mind?

We take for granted the unity of our conscious experience, but it is extremely difficult to account for. Our prospector's brain would have received sensory information from his ears, nose, skin and muscles. Dedicated centres of his cerebral cortex (somatosensory cortex, auditory cortex, visual cortex etc.) would have processed that information. His memories would have been held somewhere else, perhaps in the temporal lobe; the calculations he made on the value of gold or the cost of his supplies might have been performed in his frontal lobe. Even a single sensory input, such as his vision, would have been processed in different areas of the visual cortex. The man might see a grey rock tumbling down the slope, but the greyness would have been encoded in one area, the shape of the rock in another, its texture in another, its motion in yet another. But the man did not see: grey + round + rough + moving; he saw a rock tumbling down the slope. How did the prospector's brain integrate all this diverse information into a single conscious experience?

Consciousness appears to be *parallel,* in the sense that we can be aware of many items at once (think of how much information is contained within a single visual field),[7] but serial in the sense that we have just one stream of consciousness (we can't think two thoughts simultaneously). How does this *serial parallelism* work? Can a machine's *mind* similarly monitor parallel streams of information? Consider your television which, depending on where you live and the kind of receiver you have, may be able to receive signals from one to several hundred channels. However, unlike your brain, your television can be tuned to only one channel at a time. Even the complex image projected onto the television screen is something of an illusion. In reality the TV processes only a single signal (equivalent to a zero or one) at any moment in time, and thereby fires (or does not fire) a stream of electrons at a particular spot on the screen. It *paints* the image on the screen by performing this action thousands of times a second; and the relatively long duration of the screen's consequent scintillation does the rest. But if we could ask the *brain* of our TV set what it was *looking at*, at any moment, it would describe just a single dot.

A slightly more realistic model of conscious brain activity would be a group of five sensory devices (video camera, microphone etc., corresponding to the five senses) that record different aspects of the external world and feed their signals into a computer for analysis. But this leaves us with five independent streams of information to be processed by five independent computers. To integrate the parallel streams of information, we might use a parallel computer. A normal desktop PC is likely to have only a single linear processor, but parallel computers have many processors that are capable of performing multiple calculations simultaneously. The brain of IBM's RS/6000 'DeeperBlue' supercomputer, which defeated the great chess grandmaster, Gary Kasparov in 1998, was built from thirty-two parallel processors which independently performed the various computational tasks associated with calculating each chess move. We might similarly integrate the information from all the sensory devices by digitizing their signals and feeding them into a parallel computer. We could program our parallel computer to perform a certain action whenever it received a certain combination of stimuli from its sense organs. To give the computer a little more character, let us install it into a robot – we will call it 'Gold Digger Mark I' – and program it to march whenever he *saw* an image of the Klondike Pass on its video channel and *heard* the sound of howling winds from its microphone. Is this how our gold prospector's brain made his decision whether or not to climb the mountain?

This question is harder because, on the face of it, Gold Digger Mark I would be able to make the same kind of decisions as the prospector. But this is to ignore the subjective experience of our consciousness, which appears very different from the machinations of even a parallel computer. Parallel computers aren't really parallel in the way our consciousness appears to be. When the Deeper Blue supercomputer *thought*, each parallel stream of information from the independent linear processors was fed (as a single linear sequence of binary digits) into a (serial) controlling processor. This central processor looked at each input in turn, and performed a calculation (algorithmic routine) to transform that input into a number (to be stored in its memory), before turning to the next information stream. In reality, parallel computers are nothing more than a bunch of serial computers strapped together with another serial computer on top to integrate the streams of information.

Gold Digger's brain would similarly be *aware* of only one stream of binary information. But the prospector was not aware of one rock on the mountain, then another, then another, followed by the sound of the wind, then the temperature and so on. His conscious mind appeared to be aware of all these at once as a single integrated view of reality. What is *seeing* all this information?

Scientists and philosophers used to imagine a part of the brain that *watched* all the streams of sensory data: the *Cartesian Theatre*. Descartes even proposed a site for this theatre in the pineal gland.[8] But there is no evidence for such a privileged area in the brain and most scientists believe that consciousness is more diffusely located as part of a distributed network of neurones. We could mimic this distributed neuronal network within Gold Digger's computer console by wiring each independent processor together so that they – by their interactions – generate the final output. Gold Digger Mark II would then have what is termed a *neural net* which more closely models the connectivity of the brain. Neural nets have, of course, been built and show many interesting characteristics reminiscent of brain activity. They can, for instance, be trained to perform a difficult task such as pattern recognition. As before, we could train our Gold Digger Mark II's neural net to respond (march) whenever its video camera saw a mountain and its microphone recorded the sound of howling winds. Does the neural net's *awareness* mimic our conscious awareness of parallel streams of information in the brain?

Many neuroscientists believe that it does – that consciousness is a by-product of the brain's fantastic level of neural net interconnectivity. For instance, in the words of the neuroscientist Marcel Kinsbourne: 'Being conscious is what it is like to have neural circuitry in particular interactive functional states.'[9] The problem with this explanation is again: why should it be conscious? We know that much of the complicated work our brains perform never makes it to our consciousness. For instance, a violinist playing directly from a musical score will perform the complex neural calculations required to direct her hand, arm and upper body movement, without being conscious of this dense mass of calculation. Yet tap that same violinist on the shoulder whilst playing and she will become acutely aware of your interruption. What is the difference between complex neural nets that are conscious (that register the tap) and those that may be equally or even more complex (those that direct playing of the violin) but are unconscious?

It is hard to dispel the impression that consciousness represents an altogether different kind of operation from the one that drives unconscious actions. Most of the time I drive my car more or less *unconsciously*, allowing my unconscious mind to perform all the necessary calculations concerned with turning the wheel or depressing the brake to follow the road's twists and turns. I am not really aware of these actions; I might be listening to the radio, or thinking about some problem at work. However, if I happen to spot a hazard sign in the road – perhaps SLIP-PERY ROAD AHEAD – my conscious mind will seem to *take control* of driving the car. The radio will be forgotten and my conscious mind will instead take over the task of moving my limbs. What is *taking control* in these situations?

There are many *explanations* of consciousness and it would take several volumes to do them justice. I refer the interested reader to the many excellent books that give the theories a fairer hearing.[10] However, in my opinion, none offers an explanation that adequately accounts for the fundamental problems of consciousness: what is *awareness*? how is our (apparently serial) mind aware of so many things at once? and how do we will actions? One of the most intriguing explanations of consciousness that has appeared in recent years – and one with obvious relevance to this book – is that consciousness is a quantum-mechanical phenomenon.

THE QUANTUM MIND

Oxford mathematician and physicist Roger Penrose proposed in *The Emperor's New Mind* (1989) that the mind is a quantum-mechanical phenomenon. Penrose believes that the phenomenon of conscious action is intimately tied up with that great mystery of quantum mechanics: the *reduction of the wave function*, discussed in earlier chapters. Many other scientists have also opted for a quantum theory of consciousness. In her book, *The Quantum Self*, American scientific philosopher Danah Zohar presented a case for a quantum-mechanical holistic psychology. Zohar's husband, Ian Marshall, proposed that the physical reality of consciousness was a neuronal Bose-Einstein condensate in the brain. More recently, Scottish chemist Graham Cairns-Smith (famous for his proposal that life originated in replicating clay minerals) took up this idea in *Evolving the Mind*. And, as mentioned in Chapter Ten, Anwit Goswami and Dennis Todd proposed that adaptive mutations and conscious volition have a common quantum-mechanical source.

There are many aspects of quantum mechanics which are attractive as an explanation of consciousness. The indeterminism of quantum measurement affords us a means of escape from Newtonian determinism – perhaps a place for our free will. In the words of Edward Teller, the Hungarian-born physicist and inventor of the hydrogen bomb, 'According to quantum mechanics we cannot exclude the possibility that free will is a part of the process by which the future is created.'[11] Quantum coherence may also help to overcome the binding problem by entangling diverse information into a single coherent quantum system. Many physicists, such as Eugene Wigner (see Chapter Nine), had already recruited consciousness as a collapsing agent in quantum measurement. If consciousness can explain quantum mechanics then perhaps quantum mechanics can explain consciousness. Furthermore allowing quantum mechanics into the brain opens up the intriguing possibility that the brain may, in fact, be a *quantum computer*.

In 1982 Richard Feynman first considered the possibility of computing with quantum objects. However, it wasn't until David Deutsch, of Oxford University, showed that a quantum computer was feasible, that the field of quantum computing really took off. The unit of information of a quantum computer, the *qubit*, is like a conventional computer bit but instead of having to be in a single state at one time (either ON or OFF), the qubit can exist as a quantum superposition of both ON and OFF simultaneously. This quantum parallelism potentially allows quantum computers to perform multiple algorithmic tasks simultaneously. A quantum computer could solve in seconds problems that would tax a conventional computer for many years.

But if quantum computers are so wonderful, why don't we all have them? The reason is that they are extraordinarily difficult to build. The problem is decoherence. Quantum computers have to remain coherent long enough to perform a calculation and to report the answer. Yet, as described earlier, quantum coherence is difficult to maintain for complex systems (like computers or brains) because the quantum particles inevitably become entangled with their environment. At the time of writing, scientists have just managed to construct quantum computers with a two-qubit brain – consisting of a pair of beryllium atoms cooled to temperatures a whisker away from absolute zero. We are still a long way from a working quantum computer.

But, is it possible that our brains are at this moment performing the kind of computational activity which has eluded many of our most brilliant scientists for over a decade? Yes it is. There are many precedents for nature discovering a technology well before man's inventions (take, for instance, flight). But if the brain is a quantum computer, then what are its qubits, its units of quantum information? Neurones are generally accepted as the units of brain information, but do not look credible candidates for quantum systems. Each neurone-firing involves the motion of billions of particles in a highly complex environment. The massive levels of environmental entanglement this must entail would (almost certainly) cause very rapid decoherence. It is very doubtful that a neurone could exist as a quantum superposition long enough to perform quantum computation.

Douglas Hameroff and Roger Penrose have proposed that the microfilaments inside neurones may instead be the qubits of our quantum brains. We have already met microfilaments in Chapter Five, as the actin tramlines on which the myosin motor protein travels. Neurones also have actin microfilaments and also slightly thicker filaments (microtubules) made up of long strings of tubulin. Like all proteins, tubulin has an electrical dipole (an asymmetric charge distribution – see Chapter Five); and can exist in several conformational states. Hameroff proposed that flipping between conformational states causes electrical disturbances that propagate along the microtubules to transmit information. Penrose and Hameroff went on to propose that these electrical excitations may cause coherent oscillations within and between neurones, thereby acting as the qubits of a neuronal quantum computer. In their view, it is microtubules, rather than neurones, that represent the brain's fundamental computational unit.

I remain unconvinced by microtubules' proposed role in neuronal computing. They do not appear either sufficiently isolated or stable to remain quantum coherent. Microtubules have well-defined roles in neurones; they are tramlines for the transport of material (such as neurotransmitter) along the axon. Kinesin, a biochemical motor – a bit like the myosin motor – runs up and down the microtubules carrying vesicles filled with neurotransmitter to the synaptic knob. The microtubules are also in a constant state of flux: the tubulin protein units continually polymerizing and depolymerizing in response to changes in the cell's biochemical environment. Maintaining quantum coherence along and between all these busy structures would be the neurobiological equivalent of walking on water.

THE CONSCIOUS FIELD

There is, however, a perfectly good wave mechanical system in the brain: the electromagnetic field (em-field). All electrical phenomena involve the generation of electromagnetic fields. Neurones have massive voltage differences across their cell-membrane and voltage is, of course, a measure of the em-field's gradient. But this field will extend beyond the neurone. The fields generated by one hundred billion neurones must overlap and superimpose, to generate an extraordinarily complex em-field inside our brain.

The dynamics of electromagnetic fields are always wave mechanical. Light waves are an oscillation of the electromagnetic field, displaying all the quantum-mechanical phenomena of interference (the two-slit experiment), superpositions (the Polaroid lens experiment), and uncertainty at any temperature. It is only matter, made up of atoms and molecules, which generally hides its *waviness* beneath a cloak of decoherence at normal temperatures.

The philosopher, Karl Popper, proposed in 1993 that consciousness was a manifestation of some kind of force field in the brain and this idea was further developed and extended by Lindahl and (asA)rhem (1994). Popper pointed out that many of the properties of mind were also properties of forces (mind is incorporeal yet capable of being influenced by matter and of influencing matter in turn – so are forces). He proposed that the mind is a three-layered structure. The neurones with their action potentials represent the bottom layer that interacts directly with the body. The next layer, the 'electromagnetic wave fields (produced by neural activities) . . . represents the unconscious part of our mind.' This unconscious field would interact with neuronal activity through the forces it generates. Lastly, the 'conscious mind – our conscious mental intensities, our conscious experiences – is capable of interacting with these unconscious physical force fields'.

Popper's suggestion resonates strongly with my own subjective experience of consciousness. The representation of thoughts and ideas as waves ebbing and flowing throughout the brain seems to describe my state of consciousness far better than any neuronal-firing model. However, Popper's proposal still leaves our conscious minds somewhere out there – in the third layer – not really part of the physical brain although communicating with it through the (unconscious) em-field. What this conscious layer is made up of, and how it communicates with the unconscious em-field, is left undefined.

The neurobiologist Benjamin Libet (who performed the neuronal initiation experiments described) proposed an alternative field theory of mind with two, rather than Popper's three, layers. The brain with its action potentials still represents the bottom layer, but above this is the *conscious mental field* (CMF) that generates 'a unified or unitary subjective experience'. The CMF would have a 'causal ability to affect or alter neuronal function' and thereby provide the veto or reinforcing role on unconsciously initiated actions, which Libet proposed for his volition experiment. Libet's CMF is more economical than Popper's model; but its nature remains mysterious. Libet states that the CMF 'would *not* be a category of known physical fields, such as electromagnetic, gravitational, etc. The conscious mental field would be in a phenomenologically independent category; it is not describable in terms of any externally observable physical events or any known physical theory as presently constituted.' However, a field affected by the electrical activity in the brain, which is in turn able to modify that electrical activity, seems to me virtually indistinguishable from the brain's conventional electromagnetic field. Rigorous application of Occam's razor would leave just a one entity: the conscious electromagnetic or the Cem-field.

All electrical activity induces an em-field (as in a radio transmitter) and the induced field modifies that electrical activity (as in a radio receiver). Neuronal electrical activity in the brain will induce an em-field, and that field must, in turn, modify neural electrical activity (whether it causes changes in firing patterns is a more difficult problem that I will be returning to). It makes much more physical sense to me to simply equate the conscious mental field with the induced em-field of the brain: the Cem-field.

It may seem peculiar to ascribe the reality of our thoughts to something as ephemeral as an electromagnetic field, but it isn't. We are impressed with matter as representing the ultimate corporeal reality but it is no more *real* than radiation. Einstein's famous equation ($E = mc^2$) tells us that matter and energy are two manifestations of the same thing: a kind of matter-energy. Indeed, our exploration of the source of motion in Chapter Six, demonstrated that all the interactions we see between the objects around us (such as kicking a football) are really conducted through em-fields. It is our boot's electromagnetic field, not the boot itself that moves the football. So why can't the thought, 'kick,' be an em-field within our brain, initiating the neuronal firing that leads to that kick?

The concept of information encoded within em-fields is also a very familiar one to us. Most of my thoughts seem composed of words and images, but this kind of visual and auditory information is routinely transmitted through space to our TV screens by em-fields. When our receiver picks up the waves, they are converted to electrical activity to make the sound and pictures on the screen. Similarly, our brain may be the receiver that picks up the auditory and visual information held within the em-field of our conscious thoughts. When we think apple, the concept *apple* may be held in our brain – not as a specific neuronal-firing pattern – but as a complex em wave induced by the firing of many neurones concerned with its colour, shape, texture, etc. Each neurone contributing to the thought will generate its own em-field but these will superimpose – with appropriate reinforcements and interferences – to form the complex wave corresponding to apple inside our mind.

But is there any evidence for this? It may all sound somewhat far-fetched but it only requires three propositions to be true. The first is that our brain generates an em-field encompassing a significant fraction of its neurones. The second is that our consciousness is a product of our brain's em-field. The third is that the conscious em-field of the brain influences neuronal firing. If each of these is shown to be true, a conscious em-field is inevitable. Fortunately, they are all testable.

BRAIN WAVES

The existence of an em-field associated with the brain was known as far back as 1875, when the English physiologist Richard Canton made electrical recordings from the brain surface of dogs and rabbits. Today, electroencephalogram (EEG) monitoring is routinely performed by harmlessly placing electrodes on the skin surface above the subject's skull, to record em waves generated by electrical activity in the outer surface (the cerebral cortex) of the brain. The characteristic rhythms (*alpha, beta, theta* and *delta*) vary with the subject's alertness, yet their source is still mysterious. We know the firing of individual neurones cannot be generating them. The signal from any single neurone would be far too weak to be detected. The waves must, instead, be a manifestation of the synchronous firing of many thousands of neurones from different regions of the cerebral cortex.

It is unlikely that the physical reality of our consciousness could be the em-field that encompasses the whole brain. Patients who have had large parts of their cortex destroyed often remain fully conscious. A famous case was that of Phineas Gage, who was foreman of a New England railway construction gang, when in 1848 an accidental explosion shot an iron bar (three feet long and over an inch thick) through his left eye socket, up into his frontal lobe, and out through the top of his skull. The bar took with it a big chunk of Mr Gage's frontal lobe, yet he remained conscious and recovered well enough to return to work some seven months later. He did not however retain his job, as his personality had drastically changed. A Boston physician named Harlow described Gage as 'fitful, irreverent, indulging in the grossest profanity'. But he also noted that 'The now extremely rude Phineas Gage is an object of immense medical interest, for it seems clear, from his somewhat crude experience of psychosurgery, that one can alter the social behaviour of the human animal by physically interfering with the frontal lobes of the brain.' Mr Gage died fifteen years later, but Dr Harlow's observation became one of the inspirations that led to the infamous (and now discredited) practice of performing frontal lobotomies on psychiatric patients.[12]

So we cannot equate consciousness with any field that overarches the entire brain. Instead the em-field of consciousness is likely to be much more localized within our brain, encompassing many millions of neurones within the cerebral cortex and thalamus regions, but its precise location may shift and change in response to changing neuronal activity. Scanning techniques such as electroencephalogram (EEG) or magneto-encephalogram (MEG) are used to detect these shifts and changes in the brain's em-field. Event-related potentials (ERPs) of the order of tens of volts per metre (voltage is a measure of the gradient of the electric field) are generated in response to a variety of auditory, visual and tactile stimuli.[13]

The brain's conscious em-field must also be relatively robust since it should not be significantly affected by the electromagnetic fields encountered in daily life (although whether we could *know* our thoughts were being modified by external fields is a difficult question: *whose* mind would know?). However, this is not such a problem as may at first appear. Movement of electrical charges in the head neutralizes external electric fields to form what is known as a 'Faraday cage' that protects the brain from most of the electrical fields we meet. We are however relatively transparent to magnetic fields, and patients undergoing magnetic resonance imaging (MRI) scanning are routinely exposed to very strong magnetic fields. The MRI field will inevitably remodel the magnetic component of the (proposed) conscious em-field in the brains of patients undergoing scanning. Yet there is no evidence that MRI scanning causes any significant changes to our thoughts or actions (none at least that can be distinguished from those provoked by load banging generated by the electric coils). However for any modulation of the Cem-field to have an observable effect, it must modify nerve-firing patterns. The static magnetic fields employed in MRI scanning, couple only very weakly with tissue and are unlikely to significantly affect neurones. Changing magnetic fields couple more strongly with tissue by inducing electrical fields that may stimulate neurone firing. And there is abundant evidence (see below) that rapidly changing magnetic fields do indeed affect brain activity.

DANCING TO THE SAME FIDDLE

The second proposition, that our conscious mind is a component of the em-field is far trickier to prove, particularly since nobody can agree on what consciousness actually is. Libet has proposed a curious test for his CMF theory of human consciousness which could work equally well for the Cem-field theory. It would, however, involve some rather tricky neurosurgery. Libet suggested that during therapeutic excision of a portion of the cortex, a slab of its tissue be kept alive for experimentation. The excised brain tissue would be placed back *in situ* inside the brain, but with all its neuronal connections severed. If fields are involved in consciousness, then the excised tissue field may still be able to interact with that of healthy tissue and impact on the subject's conscious experience. If, for instance, the tissue was from the visual cortex, electrical stimulation might cause the subject to see lights despite the fact that he was no longer hard-wired to the part of his brain being stimulated.

Whether such an experiment would be practically (or ethically) feasible is one I happily leave to neurosurgeons. But there may be easier ways to test whether the physical basis of consciousness is the Cem-field. A prediction of the theory is that conscious awareness should correlate with changes to the Cem-field. The simplest way for neuronal activity to impact on the em-field is for many neurones to fire; and there is abundant evidence that this is indeed a factor in conscious awareness. However, in itself, this does not distinguish between a neuronal and a field theory of consciousness. But recall that a field is made up of waves with all the interference effects discussed in earlier chapters. If many neurones fired randomly, the peaks and troughs of their individual EM waves would not coincide but interfere, generating a zero net field (or to put it another way: the waves would decohere). For neuronal firing to have a powerful impact on the conscious field, neurones must fire in synchrony – so that their em-fields' peaks and troughs will march in step, reinforcing one another.

Reinhard Eckhorn and colleagues at Marburg's Philipps University, and Wolf Singer and colleagues at Frankfurt's Max Plank Institute for Brain Research discovered that when animals perceived visual stimuli, local and distant clusters of neurones in their visual cortex fired in synchrony to generate coherent 40–80 Hertz (oscillations per second) brain waves.[14] The researchers went on to suggest that these oscillations link distant neurones involved in different aspects (colour, shape, movement etc.) of the same visual perceptions, and could bind together features of a sensory stimulus by generating synchrony between discrete cortical areas. Wolf Singer's group discovered that when cats were shown two independent images of a bar moving in different directions on a screen, then individual neurones that responded to each image would fire at different times, asynchronously. However, when those same bars moved together (as a single bar), then the nerve cells fired in synchrony. It appeared that the cats registered each bar as a single pattern of neuronal firing, but their *awareness* that the bars represented two aspects of the same object was encoded by synchrony of firing.

Even more startling, were experiments performed using an arrangement of mirrors to present a different moving image to each eye. The experimenters monitored each cat's eye movements to determine which image it perceived: the assumption being that its eye would follow the image its attention was focused on. When only one image was presented, then only that image was perceived. However, by presenting a rival image to the other eye, the experimenters could *interfere* (perhaps wave interference?) with the first images perception and capture awareness. Remarkably, the awareness of an image did not generate any change in the number or frequency of neuronal firing events in the visual cortex, but did change their synchrony. When the cats focused upon a particular image, those neurones which *saw* that image, fired in synchrony. When awareness was lost, those same neurones still fired, but randomly. Once again, awareness correlated, not with a pattern of neuronal firing, but with synchrony of firing.[15]

If synchrony is important for awareness, we would expect that disrupting it would disrupt awareness. Gilles Laurent (and colleagues) at the California Institute of Technology examined this question in insects. Locusts have about a thousand neurones in their brain's antennal lobe, involved in the sense of smell. When the insects sniff a particular odour roughly a hundred of these neurones fire. However, it was not merely the pattern of neurones that seemed to carry information about odour, but the synchrony between individual neuronal firings. They also discovered that a neurotoxin (picrotoxin) abolished the synchrony of firing. They were then in a position to address whether synchronous firing actually means anything to the insects. For this purpose, they switched to honeybees which can be trained. Rather like Pavlov's dog, honeybees can be conditioned to stick out their tongues to obtain a reward when they smell a particular odour. However, when treated with picrotoxin, the bees lost the ability to discriminate between similar scents. Awareness of the difference between these scents appeared to be encoded in the synchronous firing of their neurones.

Examining the role synchronous firing plays in perception in the human brain is much more difficult since we cannot easily monitor the firing of individual neurones. There is however abundant evidence from

EEG and MEG studies that synchronous firing in different regions of the cortex (to generate an EEG wave) correlates with awareness and attention.[16] Experiments from Paris' Laboratoire de Neurosciences Cognitives et Imagerie Cérébrale and Germany's Institute of Psychology, demonstrated synchronous firing in distinct brain regions when a subject's attention is aroused.[17] In the Paris experiments, subjects were shown black and white patterns vaguely resembling a human face. When the subjects *saw* nothing but black and white patterns (they did not recognize the image as a face), their neurones fired asynchronously. But when the subjects recognized they were looking at an image of a human face, their neurones snapped into phase and fired synchronously. In the German experiments, the subjects were shown a visual stimulus – a red or green light – accompanied by a small (relatively painless) electric shock to one finger. Subjects soon learnt to associate the coloured light with the expectation of receiving a shock, and this associative learning was accompanied by synchronous firing in the cortex regions involved in the visual stimulus, together with that representing the hand receiving the stimulus.

There is also some circumstantial evidence that some anaesthetics disrupt synchronous firing: anaesthesia is certainly associated with a lack of awareness in humans. Indeed, signs of wakefulness (movement, eye-opening) in women under a general anaesthetic for caesarean section were found to be associated with the restoration of 30–40 Hertz oscillations in brain activity. Morphine has also been found to disrupt synchronous firing of neurones in rat brains, indicating that human morphine-induced hallucinations are probably also associated with the disruption to synchronous firing.[18]

How does the brain use synchrony? How does it even detect it? Many neurophysiologists consider synchrony a by-product of a process not relevant to its mechanism, rather like the whistle of a steam train; whilst others, such as Eckhorn, believe that the brain uses these phase-locked oscillations to tie separately processed features together into a single perceived object. However, it is still unclear how the brain *uses* synchronous firing to tie perception together. What part of the brain *oversees* these distant firings? The simplest explanation seems to be that synchronous firing generates coherent disturbances to the Cem-field and thereby impacts on our consciousness (I have no problem with the concept that a bee or indeed any sentient animal has some degree of consciousness. Until we know how it is encoded, I don't see how we can exclude any animal from consciousness.)

The Cem-field theory of consciousness would also predict that stimuli which do not reach our consciousness should not disturb it. This can be tested during habituation, the phenomenon that we no longer notice a particular stimulus (for instance, a clock's ticking) when it is mono- tonously repeated. Although we can't examine the Cem-field directly (since we don't yet know where it is localized in the brain or even if it *is* localized), there is abundant evidence that habituation in animals and man is accompanied by a reduction in perturbations to the brain's overall electromagnetic field. There have been numerous experiments – on both humans and animals – which have demonstrated habituation in EEG patterns: the subject's EEG response to a stimulus, such as a loud noise, diminishes when it is repeated. EEG measures the component of the brain's em-field outside the head but magnetoencephalography (MEG) can directly measure the brain's em-field inside brain tissue.[19] MEG detects perturbations to the brain's em-field when a subject perceives a visual or auditory stimulus, and various studies have demonstrated that the strength of these perturbations diminishes with habituation.

So evidence abounds that changes to the brain's em-field correlate with conscious awareness. This does not of course prove that these em- field perturbations are *our thoughts*, but it is at least consistent with that hypothesis.

WAVES MOVE MATTER

The third (and final) proposition of the Cem-field theory is that it impacts on neuronal firing and thereby *wills* our actions. Em-fields routinely modify the electric currents in our radio and television receivers, but can they similarly modify the brain's electric currents? As described above, neuronal firing is normally triggered by the opening of voltage-gated ion channels. Voltage is a measure of the difference between the electromag- netic field at two points so voltage-gated channels are sensitive to the brain's em-field.

Voltage-gated ion channels *see* the em-field because they possess charged amino acids that move in it. The channels are composed of a ring of proteins surrounding a cell-membrane pore which allows ions in and out. Each protein consists of a string of amino acids that loop in and out of the membrane. One of the loops (the S4 segment) contains a stretch of positively charged amino acids which seems to act as a lid on the pore. As we discovered in Chapter Six, charges experience a force in an em-field, so the charged protein lid will respond to changes in the em-field by moving to a position where their potential energy is minimum. This motion (or action) is thought to be responsible for the opening or closing of the pore.

The em-field in the neurone membrane will be modified by the global em-field. There is, therefore, the potential for the brain's em-field to modify neuronal firing patterns. However, the voltage difference across the cell-membrane is enormous (thousands of volts per centimetre). The voltage drop that triggers neuronal firing (from −65 to −40 millivolts) represents a shift of about 5,000 volts per centimetre – a very steep change of the em-field across the membrane.

The gradients of the global em-field are far smaller than this, so on its own, the global em-field would be insufficient to trigger neuronal firing from a resting state. However, neurophysiologists have long known that neurones exhibit a considerable range of excitability (epileptic seizures occur when neurones become uncontrollably excited). So, amongst the electronic network of one hundred billion neurones in our brian, there will be very many neurones fluctuating around the threshold potential necessary for firing. These undecided neurones will be very sensitive to the brain's em-field. Sometimes the em-field will reinforce the voltage difference across the cell membrane to stimulate neuronal firing; on other occasions, the em-field will diminish it to suppress firing.

Proving that the brain's own em-field modifies neuronal firing is difficult but abundant evidence suggests that relatively weak external electromagnetic fields can impact on neuronal activity. Slices of guinea pig and turtle brain have been shown to respond to external em-fields as low as a few volts per metre.[20] Isolated neurons can also respond to weak electric and magnetic fields.[21] The evoked potentials generated in living brains by sensory stimuli are usually stronger than the relatively weak fields used in these experiments. If neuronal firing patterns are modified by external fields then they are surely also modulated by the brain's own fields.

External fields have also been shown to affect brain activity in animals and man. Henry Lai (and colleagues) at Washington University demonstrated that rats exposed to microwave-frequency radiation were less able to find their way through a maze.[22] Work by C. K. Chou and Arthur Guy of Seattle's Neuroscience Medical Center has shown that microwave radiation can induce sensory auditory responses in rats and guinea-pigs (the animals *hear* the field);[23] and very many further studies have found that exposure to em-fields can change the patterns of neurotransmitter release in experimental animals. A number of studies in human volunteers have demonstrated that electromagnetic fields produce changes in EEG profiles, particularly during sleep; and very many studies on the effect of mobile phones or overhead electrical cables on human health and cognitive skills, though often with conflicting results.[24] A recent trial performed by Dr Alan Preece of the University of Bristol discovered that those subjected to mobile-phone-frequency microwave radiation had quicker response times than control subjects. The strength of the induced em-fields in the brains of subjects exposed to external sources of electromagnetic radiation is usually much lower than the fields generated by the brain's own activity.[25] Electromagnetic fields have even been used therapeutically. *Transcranial magnetic stimulation* of the brain by electrical coils placed on the scalp generates induced electric fields which excite cortical neurones and has been used to treat psychiatric disorders such as depression.[26]

If external em-fields can perturb the brain's neuronal firing, then it seems reasonable to conclude that the brain's own em-field may similarly modulate neuronal firing. Generated by neuronal activity, the Cem-field will loop back to influence neuronal firing and thereby be capable of consciously *willing* our actions. This feedback loop provides the kind of self-referral that many cognitive scientists and philosophers believe to be crucial to consciousness.

A CONSCIOUS COMPUTER

With our Cem-field theory of consciousness, let us make further modifications to the Gold Digger Mark II robot to give him a semblance of Cem-field consciousness. The first ingredient is already there: the em-field of his brain circuitry. If this em-field overarched his brain's entire circuitry (whether a parallel computer or a neural net), the field would integrate information from all calculations performed by all his logic gates. The em-field would then have some characteristics of consciousness: we could hypothesize that this field would be *aware* of the (neuronal) electrical activity generating it. However, most importantly, there would be no way to test this hypothesis since his em-field, as it stands, would be dumb. There is no way such a field could report its state to us. Gold Digger couldn't tell us whether he was conscious or not.

To have a voice, Gold Digger's em-field must be more than *aware*: it must communicate. We could engineer a communication channel by copying our own brain's architecture and installing some em-sensitive logic gates. The computational processes would then loop back upon themselves, through the electromagnetic field and the em-sensitive logic gates, to influence their own computation processes and generate an em-field sensitive output. The em-field sensitive circuitry could drive a voice synthesizer to give Gold Digger Mark III's em-field an audible voice. We could program Gold Digger to speak whenever his electromagnetic field contained visual information corresponding to the Klondike (from his video camera), together with howling winds (from his microphone) and to say, 'I see a mountain and it is cold and windy.' The electrical activity that generated speech would feedback in turn into the em-field so that Gold Digger's em-field would become em-field-aware of his action in speaking. We could program him to report on this awareness (whenever the electrical activity corresponding to initiating speech became components of his Cem-field) by saying, 'I am aware that I have spoken of the Klondike.' And who could say he was lying?

With even more sophisticated programming, we could engineer Gold Digger to perform continuous analysis of his em-field's contents (generated by both his sensory input and motor outputs) and describe them to us in a stream of consciousness report. Unlike Mark II, Gold Digger Mark III would be instantaneously *aware* of all his sensory information as a single Cem-field. Integrating his em-field sensitive circuitry with the em-insensitive classical computational process would allow the robot to work on two levels. The first would be an unconscious serial or classically parallel computation able to perform routine tasks (general electrical and mechanical maintenance) as well as driving the walking machinery and maintaining his balance – tasks best handled by classical computational number-crunching. The second level would be that of his em-field sensitive circuitry, which would receive the same sensory input as the unconscious part of his brain but function on a wave-mechanical level. These circuits would drive Gold Digger's voice synthesizer but would also have the ability to interrupt some lower-level computations to make him stop, start or change direction. We could engineer this high-level override to take over Gold Digger's motor actions whenever a certain combination of input (image of the Klondike *plus* howling wind) entered his em-field. Gold Digger would be *aware* of these *voluntary actions* since they would instantly feed back into his own em-field.

It is of course unreasonable to propose that Gold Digger Mark III, constructed with current computing technology, would have anything other than a very rudimentary consciousness. His em-field could certainly not compete with the complexity of a Cem-field generated by a significant portion of the 10^{11} brain neurones. But a computer brain constructed with this em-field feedback-loop would, I believe, possess some primitive form of consciousness, perhaps equivalent to that of animals with simple nervous systems.

Imagine now a biological version of Gold Digger's brain (switching back to neuronal circuitry) in a primitive animal. Since the brain's em-field modifies neuronal firing, it must affect some aspects of the animal's behaviour. This em-field will inevitably become subject to natural selection. The field's ability to instantly process information from millions of spatially separated neurones would surely be harnessed by evolution. Over millions of years, natural selection will inevitably modify the brain's em-field's organization and dynamics, optimizing its interaction with the neuronal network. Conversely, other neuronal circuits, needing insensitivity to the em-field's vagaries (such as, for instance, those controlling general movement or body temperature) would be insulated, protecting their computations from the em-field. The animal's brain would diverge into a robust *unconscious* number-crunching neuronal network that would take over all the brain's automated tasks, and a conscious wave-mechanical system performing *voluntary* actions. In short, the system would evolve into conscious minds.

This Cem-field theory of consciousness gives a physical reality to the powerful perception of dualism inside our own minds. The reason why it feels like our conscious mind takes over when we are driving and suddenly spot a hazard sign, is that our conscious mind *does* take over. At these points the conscious em-field – able to integrate complex information much more rapidly than the neuronal number-crunching network – overrides the neuronal circuitry to make *voluntary* actions. This theory restores a measure of dualism to the mind; but a dualism rooted in physical reality. One part of the mind – the unconscious – is matter-based; the other – the conscious mind – is an energy field. Both aspects of the mind are equally *real*; they just have different physical manifestations.

But, you may well say, the neurones involved in unconscious brain activity must also have an em-field. Why aren't these fields conscious also? Furthermore, why isn't my television set, which also generates an em-field, conscious? The somewhat surprising answer is that we have no way of knowing whether or not any of these fields are indeed conscious. The only conscious minds that can state that they are conscious are those that can communicate their consciousness. That information could be in the form of speech, sign language or words on a VDU screen, it could even be encoded in generation of a particular odour (remembering Samuel Beckett's 'I stink therefore I am'). But to be demonstrably conscious, it must communicate.

There is evidence that, in some circumstances, parts of the brain may be conscious, but are unable or have only very limited abilities to communicate. Roger Sperry and Ronald Meyers discovered the 'split brain' phenomenon in their animal experiments in the late 1950s. In the 1960s, patients suffering from severe epilepsy who did not respond to conventional treatments were subjected to the drastic treatment of severing the *corpus callosum* in their brains. The *corpus callosum* is a bundle of nerve fibres connecting the brain's left and right hemispheres, which communicates information between them. You may know that, with a few exceptions, the brain's left and right hemispheres receive sensory information from, and control, the opposite halves of the body. For example, your left hemisphere controls the movement of your right hand; your right hemisphere receives sensory information from the left-side of your visual field. However the brain's centre for speech interpretation and production is located in only one hemisphere: the left.

The split-brain patients appeared perfectly normal and their seizures were gone. They could talk and read and seemed happy, alert and healthy. Yet Sperry discovered that they had a startling deficit. In one experiment, a word (for example 'fork') was flashed up so only a patient's right hemisphere could receive the information. They would not be able to say what the word was. However, if the subject was asked to write what he saw, his left hand (controlled by his right hemisphere) would write 'fork'. If asked what he had written, the patient would have no idea. His talking (left-hemisphere mind) would be completely unaware of what his mute (right-hemisphere mind) was up to. He would know he had written something, yet could not tell observers what the word was. Similarly, if the patient was blindfolded and some familiar object, such as a toothbrush, was placed in his left hand, he appeared to know what it was – for example by making the gesture of brushing his teeth – yet would be unable to name the object. But if the left hand passed the toothbrush to the right hand, the patient would immediately say 'toothbrush'.

Whether these patients' right hemisphere was conscious – aware of what it was doing – is impossible to say. Lacking speech, the right hemisphere was unable to say whether it was conscious or not. The brain's right hemisphere may, on these grounds, be an automaton brain but it could equally be considered to be a conscious but mute mind. Similarly, there may be distinct em-fields in intact brains separated from the one that we – as speaking people – are aware of. The only conscious minds that we are able to listen to, are those that can talk.

So the conscious em-field must inevitably be located in those brain areas which influence motor-neurone firing sufficiently to communicate: the motor, sensory and visual cortex, the centres concerned with speech and the temporal lobes connected to memory. People with intact brains are conscious of the neural activities of both halves, because these are communicated to the speaking part through the *corpus callosum*. Once that link is severed, the right hemisphere is mute and its consciousness becomes purely philosophical. Similarly, whether the brain's other non-speaking regions are conscious, or indeed any other em-fields are conscious, are unanswerable questions.

I strongly suspect, however, that there is only one consciousness in our brains and that inanimate electrical devices are not conscious. I believe that consciousness is not just any old electromagnetic field. Just as not all matter is living, not all em-fields are conscious. Our conscious minds have been modified and improved over millions of years of evolution to perform the function of conscious-decision making. A mute and powerless em-field would have no function, and thus could not have contributed to its host's fitness. Without evolutionary development, it would be left as a disorganized *primordial* field with only the faintest semblance of consciousness.

The great advantage of the Cem-field as a theory of consciousness is that it is simple, and makes testable predictions. It involves no new physics or biology. All that is required is a straightforward and inevitable feedback loop between the brain's neuronal network, and the field generated by that network. This theory also has many interesting implications for our understanding of awareness, emotion, creativity, problem-solving and consciousness in animals. There are also fascinating possibilities for building and using electronic devices able to interact directly with the Cem-field.

But we must free our gold prospector from his predicament. He is still standing at the foot of the mountain, sensory data streaming into his brain neurones. His brain's neuronal network is busy performing classical algorithmic computations on the various possibilities for action; while his Cem-field (his conscious mind) is also receiving the same data, through the field induced by neuronal firing. In many cases, the stimulatory and inhibitory synaptic signals received by the decisive neurone will be sufficiently positive or negative to resolutely trigger or inhibit firing, irrespective of the Cem-field. In these circumstances, the Cem-field has no influence on the neuronal computations process, and unconscious decision-making will ensue. But in other situations the stimulatory and inhibitory inputs into the decision-making neurone(s) will not be decisive, and the neurone(s) will be left poised on the brink of an action potential. In these cases, the pushes and shoves from the Cem-field may be decisive and a conscious decision made. Under these circumstances, there will be only very small changes of energy involved in the Cem-field and neurone interaction; this inevitably returns us to this book's central theme: quantum mechanics.

MAKING A QUANTUM DECISION

When hard-nosed physicists search for terms to describe wave function collapse or quantum measurement, they even hijack terms to do with volition. 'I am not going to explain how the photons "decide" whether to bounce back or go through; that is not known.'[27] Or, 'Nature *chooses* [my italics] between one or the other of them and *actually* effects some kind of reduction procedure . . .'[28] Science writers can find no better words: 'the electron is being forced by our measurement to *choose* [my italics] one course of action out of an array of possibilities.'[29] I have, of course, used the same terms myself, and even extended the analogies with cognitive processes to include descriptions of quantum superposition and the inverse quantum Zeno effect. As with their use by *real* physicists, I have been careful to deny any kind of volition in quantum systems. Yet it remains curious that the closest concepts to quantum-mechanical phenomena are not in the physical world but in our own minds. By now, I hope you can see that there may be something more to these interesting parallels than mere coincidence.

We still have our prospector's mind, stuck in a quantum quandary with the Cem-field supplying the push necessary to initiate or repress a particular course of action. But do these interactions take place at the classical or quantum-mechanical level? This will depend on the amount of electromagnetic energy involved in opening and closing neurone ion channels. The interactions between matter and em-fields can be described quantum mechanically by the theory of quantum electrodynamics (QED – largely due to Richard Feynman, which is described in his marvellous book *QED: The Strange Theory of Light and Matter*). In QED, electromagnetic forces are transmitted by photons travelling from one particle to another. Yet iron filings moving in a bar magnet's em-field do not exhibit quantum-mechanical behaviour. Why they don't, is that the force between the magnet and the filings involves the exchange of trillions of photons, and the quantum-mechanical effects are washed out by the inevitable decoherence. The interaction between the Cem-field and neurones may therefore take place either at the quantum or the classical level, depending on the number of photons involved.

We do not yet know how many photons need to be absorbed from the electromagnetic field to open a voltage-gated ion channel, but it is likely to be very small. Ion channels in biological systems more extensively characterized are known to respond to single photons. For instance, a group of the salt-loving Halobacteria (mentioned in Chapter Two) uses a protein (bacteriorhodopsin) to perform a unique form of photosynthesis. Bacteriorhodopsin forms a pore in the bacterial cell-membrane and absorbs light energy to pump protons (hydrogen ions) through this pore and out of the cell. The bacteria utilize the resulting proton gradient to synthesize ATP. It takes the absorption of just two light photons to transport a single proton across the cell-membrane, clearly an interaction that could be quantum. Interestingly, the system also has a sensory function. Halobacteria inhabit the intensely sunlit Dead Sea, where they swim away from regions of bright sunlight – to escape sunburn. They do this, by sensing strong sunlight with a related protein channel which is also photon-sensitive and transmits a signal to the bacterial flagella, telling it to swim.

So bacterial 'eye' ion channels are sensitive to single or pairs of photons. The bacteriorhodopsin channels are similar in structure to the voltage-gated channels of neurones, so the presumption is not unreasonable that similar levels of energy exchange are involved in opening these channels. In that case, the interaction between the Cem-field and neuronal ion channels in our brain may also take place at the quantum level. The Cem-field may exist as a superposition of a field that has triggered channel opening and of one that has prevented it. The channel may persist as a superposition of an open and closed channel. But these quantum states cannot persist indefinitely. At some point the quantum states must interact with a measuring device to make one or other possibility real. When will this occur?

As in previous chapters, we should look to decoherence to provide an answer. Let us first imagine that the relevant ion channel is in a resting neurone – one without a hope of firing unless thousands of channels open. If the channel remains closed, then nothing much will happen.

However, even if the channel opens, then nothing much will happen. A few ions may travel through the pore, but after only about one millisecond the channel will spontaneously close[30] Under these circumstances, there will be minimal environmental entanglement and so decoherence will be suppressed. To put it another way, the opening/closing of the channel will be *invisible* to the neurone, which will be unable to *measure* the state of the channel. The interaction between the Cem-field and the channel may therefore remain quantum.

If, instead, the voltage gate absorbing the photon is in a neurone committed to firing (thousands of gates already open), the absorption event will similarly make no macroscopic difference to the cell or to the brain (since the neurone will fire anyway) and the interaction may once again remain quantum. However, now imagine that the channel is critical in a neurone poised on the brink of an action potential. The superposition ({photon absorbed and channel open *(+/−)* photon not absorbed and channel closed}) will now become a larger entanglement: {photon absorbed and channel open and neurone fired *(+/−)* photon not absorbed and channel closed and neurone not fired}. The channel's alternative states (open or closed) will be associated with very different fates for the neurone: firing or not firing. This quantum event will now *make a difference* to the neurone, the brain and, potentially, the life of the brain's owner. Under these circumstances (of maximum environmental entanglement), decoherence will be instantaneous. At this point the photon, as a quantum component of the Cem-field, must make a *choice* − to be absorbed or not − and a quantum measurement will be made.

At these decisive junctures, the photons that make up the Cem-field will be subject to the same kind of conditional quantum measurement I highlighted in previous chapters. The brain's neurone network and their trillions of em-field sensitive ion channels, will act as a quantum-measuring device to collapse the Cem-field quantum states, but only when it makes a difference in terms of neuronal firing. Poised on the brink of an action potential, quantum measurement may make *decisions* to perform directed actions and provide us with what we call our free will.

The Cem-field will roam through the neuronal pathways of the prospector's brain nudging and twitching various neurones; but these will remain at the quantum level unless they actually trigger neuronal firing, making a decision. Many of these Cem-field and brain interactions will involve not only a single neurone firing but a network of neuronal firing in different brain regions, to generate a particular motor action. The network which initiates a particular action may be only one possible combination of neuronal firings amongst billions of alternative firing states. But now we are back in the familiar multidimensional quantum landscape, recognizing the power of the inverse quantum Zeno effect to pave a path of quantum measurement towards a particular action.

However, I'm sure your Cem-field has had its fill of photons and ion gates so let us rush towards our happy ending. The components of the Cem-field which lit up a path of quantum measurement in our gold prospector's brain were images of his wife and children with happy faces. It is these that crashed his Cem-field out of its superposition of indecision states and led his mind along a photon-collapsing path towards a decision. That decision fired the decisive neurone that propelled him up the mountain and onwards to the Klondike. Not only did he survive but he struck gold and returned home to his wife and family with a fortune. What makes the story even more heart-warming is that he made his own decision. His conscious mind had a role to play in his actions.

Man is not an automaton. Our conscious electromagnetic field exploits quantum measurement to move particles inside our brain, providing us with the phenomenon we call free will. Consciousness *drives* free will. This quantum level control – a control lacking in robots – gives us the edge in our interactions with the world outside. It propels men and women to drag tons of supplies up frozen mountainsides. It may sometimes (at a more primitive level) be the driving force which causes a bird to soar or a salmon to leap up a waterfall. I believe it also lies at the heart of that most extraordinary of human abilities: creative thinking. Great ideas are not pulled out of the air; but out of the quantum multiverse. In a sense, our minds have recaptured the same quantum evolutionary process I believe propelled life from its origin billions of years ago and drove the evolution of living organisms towards increasing complexity. Although that process may survive and prosper inside microbes, its influence on the lives of multicellular creatures may now be buried deep within our bodies, or restricted to its negative effects such as infectious disease and cancer. Yet, by nurturing sensitivity to the brain's electromagnetic field, animals, and particularly human beings, have recaptured entanglement with a quantum-mechanical entity – the conscious mind – and once again harnessed quantum measurement to perform directed actions. We have quantum-evolved our own free will.

It has been a long way from our first sighting of the rock pigeon in flight. We have explored the extent and the limits of life and looked right into the core of living cells, uncovering their dynamics. Our search has taken us from the chaos of thermodynamics to the strangely structured world of quantum mechanics. We have examined how internal quantum measurement uniquely defines life and directs our actions. Quantum measurement may well have precipitated self-replicators out of the primordial soup, guiding their progression towards the emergence of the first living cell. Our own cells straddle two worlds: the quantum realm of fundamental particles and superposition, and the classical world of actions. This is what makes life special and powerfully different from the inanimate world. This is how consciousness endows us with free will. Life and consciousness are contingent upon the dynamics of fundamental particles. Life and consciousness are quantum phenomena.

NOTES

1: What is Life?

1. *Ancient Near Eastern Texts Relating to the Old Testament* (ed.) James B. Pritchard (Princeton University Press, 1950, p. 37 ff.).
2. *Enuma elish* ('when on high') in *Larousse World Mythology* (ed.) Pierre Grimal. (1989) p. 69.
3. Aristotle *Progression of Animals* IX (trans.) E. S. Forster (1968) p. 709b10 (Harvard Heinemann, London & Cambridge MA).
4. Aristotle *Parts of Animal* IV (trans.) A. L. Peck (1968) p. 679b30 (Harvard Heinemann, London & Cambridge MA).
5. Aristotle *Metaphysica* p.1049b6.
6. Aristotle *Movement of Animals* X (trans.) E. S. Forster (1968) p. 703a30 (Harvard Heinemann, London & Cambridge MA).
7. Cloning animals – like the famous Dolly the sheep – is a new way to propagate life, but it does not make new life. Fundamentally, it is no different to the vegetative propagation of plants.
8. The game is apparently a very old one going back to the earliest classification system by Karl Linnaeus, the Swedish founder of systematics, who in his *Systema Naturae*, divided nature into three kingdoms: animal, vegetable and mineral.
9. The word itself bears the trace of vitalism; its root is the Latin *anima*, meaning life or soul.

2: The Limits of Life

1. We will be examining chemical compounds in more detail in Chapter Five, but at this stage all we need to know is that chemical formulae describe the constituents of chemical compounds. The formula of carbon dioxide indicates that each molecule has one atom of carbon and two of oxygen.
2. The chemical name *organic* does not necessarily imply *biological*. Organic compounds are compounds of carbon, but exclude simple compounds such as carbon dioxide or cyanide. Inorganic compounds are simple carbon compounds and chemicals lacking carbon.
3. Bacteria are the smallest and simplest cellular life. Mostly they are single-celled although they sometimes grow in chains or other cell aggregates. They have *prokaryotic* cells that lack the chromosome-containing nucleus found in (*eukaryotic*) plant and animal cells; instead their DNA is found free within the cell. Bacteria should not be confused with viruses which are much simpler but do not have cells and are unable to replicate autonomously.
4. Robert L Folk (1996) 'In Defense of Nannobacteria' *Science* 274 pp. 1285e–1289e.
5. More recent analysis casts increasing doubt on a biological origin for the 'nanofossils' in ALH84001. McKay's team have turned their attention to another Martian meteorite, Nakhla, which harbours much bigger microfossil-like structures.

3: Life's Biggest Action

1. Although the ancient Romans believed that mother bears actually licked their cubs (born formless) into shape – hence today's phrase.
2. J. Watson (1968) *The Double Helix* (Weidenfeld & Nicolson, London).

4: How Did We Get Here?

1. In defence of my ancestor, I must point out that in eighteenth-century Ireland, the Irish had no rights to stock or land and so perhaps could be forgiven for the occasional transgression to feed their family and friends.
2. R. L. Tatusov, E. V. Koonin & D. J. Lipman (1997) 'A genomic perspective on protein families' *Science* 278 pp. 631–637.
3. Richard Dawkins (1986) *The Blind Watchmaker* p. 9.
4. J. Cairns, J. Overbaugh & S. Millar (1988) 'The origin of mutants' *Nature* 335 pp. 142–145.
5. B. G. Hall (1998) 'Activation of the **bgl operon** by adaptive mutation' *Molecular Biology and Evolution* 15 pp. 1–5.
6. E. V. Koonin, R. L. Tatusov, & K. E. Rudd (1995) 'Sequence similarity analysis of *Escherichia coli* proteins: functional and evolutionary implications' *Proc. Natl. Acad. Sci. USA* 92 pp. 11921–11925.
7. Sedimentary rock is formed when fine grains of rock are deposited from the air or water. The grains accumulate in layer upon layer over millions of years and are compacted and cemented by pressure and minerals to form rock. Sedimentary rock often contains the fossilized bodies of living organisms that were trapped and buried during the deposition of the sedimentary material.
8. J. W. Schopf (1993) 'Microfossils of the early Archean apex chert: new evidence of the antiquity of life' *Science* 260 pp. 640–646.
9. All elements occur as a number of isotopes that differ in the number of neutrons in their nuclei. Carbon consists of two stable isotopes, one light (Carbon-12) with six neutrons and a heavier form (Carbon-13) with seven neutrons. A third isotope (Carbon-14) is radioactive and so decays quite rapidly and is only found in trace amounts. This is the isotope used for carbon dating.
10. I should say a word here about viruses. Most microbiologists accept that viruses are not alive. Firstly, they cannot self-replicate – they need a host cell to make copies of themselves. Virus particles consist of a small molecules of RNA or DNA inside a protein shell. The virus sticks to the exterior of our cells, injecting its nucleic acid inside. This carries a genetic message that says: MAKE LOTS OF COPIES OF ME. Unfortunately, our cells slavishly obey this message and makes lots of virus particles. So, viruses need a host cell to replicate; one rather neat description of them is that they are 'bits of bad news wrapped up in protein'. Viruses' second deficiency, from the point of view of life, is that on their own they have no metabolism. They don't do anything; they cannot perform any actions. Again, they rely on the host cell to perform their actions. For these and other reasons, viruses are not thought ancestral to cellular life but rather a by-product of it.
11. G. Wächtershäuser (1988) 'Before enzymes and templates: theory of surface metabolism' *Microbiol. Rev.* 52 pp. 452–484.
12. D. H. Lee, J. R. Granja, J. A. Martinez, K. Severin & M. R. Ghadri (1996) 'A self-replicating peptide' *Nature* 382 pp. 525–528.
13. There are a number of these

fundamental constants that include the speed of light, the charge on an electron, the mass of a proton and Planck's Constant.

14. A somewhat satisfying (to a biologist) consequence of this principle is that the laws of physics become subordinate to biology. If a physicist wishes to discover why the fundamental constants have their precise value, then he or she must look to living cells for an answer.

5: Life's Actions

1. The seven zeros in ten million find their way into the pH scale. pH is the negative logarithm of the hydrogen ion concentration, so neutral water has a pH of seven.
2. I. Rayment, H. M. Holden, M. Whittaker, C. B. Yohn, M. Lorenz, K. C. Holmes & R. A. Milligan (1993) 'Structure of the actin-myosin complex and its implications for muscle contraction' *Science* 261 pp. 58–65.
3. As always bacteria are biochemically more versatile than we are. They can burn their food not only with oxygen but with other inorganic compounds, such as nitrate or sulphate.
4. R. Yasuda, H. Noji, K. Kinosita & M. Yoshida (1998) 'F1-ATPase is a highly efficient molecular motor that rotates with discrete 120 degree steps' *Cell* 93 (7) pp. 1117–1124.

6: What Makes Bodies Move?

1. The scientific term, mass, should not be confused with weight, despite the fact that both are expressed in grams and kilograms. Mass describes the quantity of matter in a body and is unaffected by gravity. A man with a body mass of eighty-five kilograms would still have the same mass if he lived on the moon, despite the fact that he would weigh a lot less.

7: What is Quantum Mechanics?

1. His hobby was Egyptology and in 1819 he provided the key that allowed Champollion to decipher the hieroglyptic text carved on the Rosetta Stone. he simply counted the number of occurrences of common words like 'and' in the Greek version and found the hieroglyphic symbol that occurred the same number of times.
2. Prove this fundamental tenet of quantum mechanics in the privacy of your own home. You will need a strong light source (a laser pointer makes an excellent source) and a dark room. A white wall can act as the detector screen and a strip of aluminium foil makes an excellent impermeable screen that can be easily pierced to form holes. Shine light through holes of different size (such as those made by a nail or a fine pin) and compare the width of the image formed on the screen with that of the hole.
3. Feynman (1963) *Lectures on Physics* pp. 37–41.
4. Arndt, M., Nairz, O., Vos-Andreae, J., Keller, C., Van der Zouw, G. and Zeilinger, A. (1999) 'Wave-particle duality of C_{60} molecules? *Nature* 401: pp. 680–682.

8: Measurement and Reality

1. Wheeler & Zurich (1983) *Quantum Theory and Measurement* p. 17.
2. There is considerable technical difficulty associated with detecting photons *in flight* so the following experiment is normally performed with a beam of electrons, not photons. The experimental set-up is essentially the same – electrons are fired from an electron gun at a scintillation screen (as in a television tube) and with a slit screen in the path of the beam. As mentioned in

the previous chapter, electrons, or indeed any fundamental particle, display the same interference effects as photons in the two-slit experiment. Electrons, like photons, must be able to travel by two routes simultaneously. Electrons are however far easier to detect during their flight than photons so the following experiment is more feasible if performed with electrons. The principles are however identical in the two systems so in the *thought experiment* described in the text we will stick to the more *illuminating* photons.

3. In reality it is more complex. The wave function amplitude is a complex number and the probability is obtained by squaring the modulus of that amplitude.

4. Quoted in John Gribbin (1996) *Schrödinger's Kittens* p. 120.

5. This kind of *vector addition* is common in mechanics. A barge can be towed along a canal by another barge pulling from a twelve-o'clock position, directly ahead. However, the same *pulling power* can be achieved by two horses, towing the barge at eleven o'clock (horse on the left bank) and one o'clock (horse on the right bank).

6. In quantum mechanics these *perpendicular* states, either the state vector or the measuring device are often referred to as *orthogonal* states. Orthogonal has a wider (and deeper) meaning than the geometric relationship of perpendicularity and can describe momentum, position or energy states.

7. Wheeler & Zurich (1983) *Quantum Theory and Measurement* p. 56.

8. For example, the sum of the infinite series $1+\frac{1}{2}+\frac{1}{4}+ \ldots$ is just 2.

9. They are actually exactly the same phenomenon viewed from different frames of reference. With a mobile microscope that moves along a path, $A{\rightarrow}B{\rightarrow}C{\rightarrow}D$, we can use the inverse Zeno effect to move an atom along

that same path. However, if we jumped aboard our microscope and performed the same series of measurements then the atom doesn't seem to be moving at all. In fact our measurements are preventing its motion by the quantum Zeno effect. Conversely, if we observed someone performing the quantum Zeno effect (to freeze motion) from a *fixed* point in space, say, the sun, then both the observer, her measuring device and the measured particle, are all moving along an orbital path: the inverse Zeno effect! Galilean relativity tells us that the laws of physics are the same in all frames of reference in uniform motion, so the two phenomena become equivalent.

10. Wheeler & Zurich (1983) *Quantum Theory and Measurement* p. 44.

9: What Does It All Mean?

1. Feynman (1965) *The Character of Physical Law*. British Broadcasting Corporation, London.

2. Wheeler & Zurich (1983) *Quantum Theory and Measurement* p. 201.

3. *Ibid* p. 56.

4. *Ibid* p. 32.

5. *Ibid* p. 5.

6. *Ibid* p. 184.

7. *Ibid* p. 201.

8. *Ibid* p. 150.

9. *Ibid* p. 157.

10. *Ibid* p. 177.

11. *Ibid* p. 209.

12. *Ibid* p. vii.

13. *Ibid.* p. 48.

14. *Ibid* p. 48.

15. *Ibid* p. 8.

16. *Ibid* p. 138.

17. *Ibid* p. 142.

18. *Ibid* p. 144.

19. *Ibid* p. 43.

20. Davies and Brown (1986) *The Ghost in the Atom* p. 71.

21. Ibid p. 84.

22. Degrees of freedom are the physicist's

ways of describing the number of different ways a body can move. Degrees of freedom generally include rotational, vibrational and translation (travelling in a particular direction) movement.

23. In some circumstances it may be possible to recapture the information that has escaped into the environment due to decoherence and reconstruct the interference pattern (although this is generally very difficult), and with it, the evidence for superposition.

10: The Beginning

1. A single letter is used to represent each amino acid so that R is arginine, M is methionine, K is lysine, Q is glutamine and so on.

2. It is of course impossible to imagine a twenty-dimensional space but mathematically, it is as easy to handle as three-dimensional space. Quantum physicists are used to dealing with much higher numbers of dimensions including infinite dimensions.

3. Once again I must emphasize that by my use of the word choice, I am *not* implying that the peptide makes any conscious decision. Unfortunately, our language is ill equipped to deal with quantum mechanics, so the use of these loaded terms is inevitable if we wish to avoid unwieldy jargon.

4. J. McFadden & G. Knowles (1997) 'Escape from evolutionary stasis by transposonmediated deleterious mutations' *Journal of Theoretical Biology* 186 pp. 441–447.

11: The Quantum Cell

1. It is in fact a thermodynamic law that absolute zero, rather like the speed of light, can never quite be attained – it is impossible to cool a body to absolute zero – but very close to zero will suffice for our purposes.

2. Depending on its energy and the thickness of the box walls, it may even be able to *quantum tunnel* through the box.

3. If a system of mass m is in a superposition of two position states (modelled as two Gaussian wave packets) separated spatially by a distance Δx then the decoherence time, t_D, is defined to be: $t_D = t_R(\lambda_T/\Delta x)^2$, where $\lambda_T = \hbar\sqrt{2mk_BT}$, is the thermal de Broglie wavelength that depends only the temperature T of the surrounding environment and for a proton at 300K works out as 0.27 (asA). The relaxation time t_R, is the time taken for the wave packets to dissipate the energy difference between the coherent states.

4. What actually happens to the interference terms is that some are positive, some negative, together adding up to zero.

5. *E. coli* is short for *Escherichia coli*, a ubiquitous inhabitant of our lower gut. It mostly does us no harm, apart from occasional bouts of traveller's diarrhoea when we meet a new strain. It does however cause serious diarrhoea in infants, and is an important cause of infant mortality worldwide.

6. A. Kohen & P. J. Klinman (1998) 'Enzyme catalysis: beyond classical paradigms.' *Accounts of Chemical Research* 31 pp. 397–404.

7. Of course the critical protons and electrons in real enzymes (like beta-galactosidase) have a much higher probability of being in the right position for optimal enzyme activity. The model enzyme's fifty per cent probability is used for ease of calculation but should not be taken too literally. Yet, even if the particles inside enzymes have a very high probability of being where they are needed, there will always be some probability of finding them elsewhere. These probabilities will be subject to quantum measurement effects, as described.

12: Quantum Evolution

1. Any keen-eyed molecular biologist may spot that I am dealing with the codon sequence of the coding strand of the DNA, rather than the more conventional complementary strand. This makes it easier to discuss the encoding of the messenger RNA sequence.

2. Tautomeric forms of DNA bases are rare, so the wave function for the coding proton will reflect this by being much bigger at its normal position than the tautomeric position.

3. It is presently unclear how often natural mutations are caused by DNA base tautomerization. However, all the mechanisms that cause mutations involve the motion of fundamental particles within the DNA double helix. They all involve the dynamics of the particles that make up the double helix or the particles that interact with the double helix, and must thereby be subject to quantum measurement.

4. *Time* is not really the right word to use here, since the proton will be in a superposition of position states at all times; but it is the most convenient description of the *amplitude* for the proton being at one position rather than another.

5. None at least feasible given the current experimental evidence. There are a few theoretical possible ways by which lactose could interact with the DNA that encodes beta-galactosidase: it might for instance bind to a protein which then binds to the gene that encodes beta-galactosidase, and somehow affect its mutation rate. However, a great many experiments have been performed that rule out these trivial explanations of the phenomenon of adaptive mutations.

6. B. G. Hall (1997) 'Spontaneous point mutations occur more often when advantageous than when neutral' *Genetics* 126 pp. 5–16

7. J. McFadden & J. Al-Khalili (1999) 'A quantum-mechanical model of adaptive mutations' *Biosystems* 50 pp. 203–211

8. Michael J. Bethe (1996) *Darwin's Black Box*

9. Gene duplication may be mediated by a variety of processes, including errors made during DNA replication.

10. Note that quantum evolution does not accelerate the overall mutational process; it is limited by mutational rate in exactly the same way as standard Darwinian evolution. Genes will not accumulate any more mutations by quantum evolution than they would be expected to by standard mechanisms; but they will be more directed.

11. K. Ghanekar, A. McBride, O. Dellagostia, S. Thorne, R. Mooney & J. McFadden (1999) 'Stimulation of transposition of the *Mycobacterium tuberculosis* insertion sequence 1S6110 by exposure to a microaerobic environment' *Molecular Microbiology* 33 pp. 982–993.

13: Mind & Matter

1. A ligand is a molecule that binds a receptor.

2. The actual stimulus that triggers the release of neurotransmitter is an inflow of calcium ions into the nerve ending, through another set of voltage-gated ion channels. The voltage drop that triggers these channels to open and allow in the calcium is the arrival of the action potential at the nerve ending.

3. In fact, nerve axons may branch into a number of axon collaterals, and each may terminate in a nerve synapse.

4. There will of course be very many neurones involved with making this or any other decision; but somewhere along the line, if a decision is being made to perform a motor action,

some neurone must fire that would not otherwise fire is that motor action was not initiated. For simplicity, we will call such a neurone, the decisive neurone, but recognizing that many of its decisions may already have been made for it by upstream neurones.

5. All modern computers can be described as universal Turing machines after Alan Turing, an extraordinarily gifted mathematician and one of the founders of the science of computing technology. Turing showed that any algorithmic operation could be broken down into a set of very simple instructions that can be performed by any general computer, or universal Turing machine.

6. The zombie metaphor was introduced by Robert Kirk's article 'Zombies vs. Materialists', published in *Mind* (1974) to question the role of consciousness in human activity. The idea was used much more extensively by David Chalmers in his book, *The Conscious Mind* (1996). Could a zombie do everyday tasks as well as conscious beings? Would we be even able to tell if our friends or colleagues were in fact unconscious zombies? The question has taxed many scientists and philosophers in the consciousness field in recent years and various scenarios have been invented to 'put zombies to the test'.

7. Many cognitive scientists assert that we can only be aware of seven to nine items at once. I am not entirely convinced by their arguments (based on experiments where subjects were asked to report back on the items in their consciousness); but in any case, each of these items (e.g. pear, orange, apple, etc.) may themselves be very complex and must be encoded by many bits of information.

8. Descartes proposed the pineal gland because it is buried deep within the brain and, unlike other structures of the brain, it is not divided into left and right hemispheres.

9. Kinsbourne, M. 1993. In Bock and Marsh 1993.

10. Chris Nunn (1996); Dennett (1991); Eccles (1989); Hofstadter (1979).

11. Edward Teller (1998) *Science* 280 pp. 1200–1201.

12. David Shutts (1982) *Lobotomy: Resort to the Knife* (Van Nostrand Reinhold Company, New York).

13. L. Kaufman, Y. Okada, J. Tripp, and H. Weinberg (1984) 'Evoked neuromagnetic fields' *Ann. N. Y. Acad. Sci.* 425 pp. 722–742.

14. C. M. Gray, P. Koenig, A. K. Engel, and W. Singer (1989) 'Oscillatory responses in cat visual cortex exhibit inter-columnar synchronization which reflects global stimulus properties' *Nature* 338 pp. 334–337; R. Eckhorn, R. Bauer, W. Jordan, M. Brosch, W. Kruse, M. Munk & H. J. Reitboeck (1988) 'Coherent oscillations: a mechanism of feature linking in the visual cortex? Multiple electrode and correlation analyses in the cat' *Biol. Cyber.n* 60 pp. 121–130.

15. M. Barinaga (1998) 'Listening in on the brain' *Science* 280 pp. 376–378.

16. G. Tononi, R. Srinivasan, D. P. Russell, and G. M. Edelman (1998) 'Investigating neural correlates of conscious perception by frequency-tagged neuromagnetic responses' *Proc. Natl. Acad. Sci. USA* 95 pp. 3198–3203.

17. E. Rodriguez, N. George, J. P. Lachaux, J. Martinerie, B. Renault, and F. J. Varela (1999) 'Perception's shadow: long-distance synchronization of human brain activity' *Nature* 397 pp. 430–433; W. H. Miltner, C. Braun, M. Arnold, H. Witte, and E. Taub (1999) 'Coherence of gamma-band EEG activity as a basis for associative learning' *Nature* 397 pp. 434–436; W. Singer (1999) 'Neurobiology. Striving for coherence' *Nature* 397 pp. 391–393.

18. M. A. Whittington, R. D. Traub, H. J. Faulkner, J. G. Jefferys & K. Chettiar (1998) 'Morphine disrupts long-range synchrony of gamma oscillations in

hippocampal slices' *Proc. Natl. Acad. Sci. USA* 95 pp. 5807–5811.

19. W. Hulstijn (1978) 'Habituation of the orienting response as a function of arousal induced by three different tasks' *Biol. Psychol.* 7 pp. 109–124; B. Rockstroh, M. Johnen, T. Elbert, W. Lutzenberger, N. Birbaumer, K. Rudolph, J. Ostwald & H. U. Schnitzler (1987) 'The pattern and habituation of the orienting response in man and rats' *Int. J. Neurosci* 37 pp. 169–182; C. Hirano, A. T. Russell, E. M. Omitz & M. Liu (1996) 'Habituation of P300 and reflex motor (startle blink) responses to repetitive startling stimuli in children' *Int. J. Psychophysiol.* 22 pp. 97–109; A. Amochaev, A. Salamy, W. Alvarez & H. Peeke (1989) 'Topographic mapping and habituation of event related EEG alpha band desynchronization' *Int. J. Neurosci.* 49 pp. 151–155; M. Johnen & H. U. Schnitzler (1989) 'Effects of a change of tone frequency on the habituated orientating response of the sleeping rat' *Psychophysiology* 26 pp. 343–351; R. N. Leaton & W. P. Jordan (1978) 'Habituation of the EEG arousal response in rats: short and long-term, frequency specificity and wake-sleep transfer' *J. Comp. Physiol. Psychol.* 92 pp. 803–814.

20. J. G. R. Jefferys (1981) 'Influence of electric fields on the excitability of granule cells in guinea-pig hippocampal slices' *Journal of Physiology* (Lond) 319 pp. 143–152; C. Y. Chan, J. Hounsgaard, and C. Nicholson (1988) 'Effects of electric fields on transmembrane potential and excitability of turtle cerebellar Purkinje cells in vitro.' *J Physiol* (Lond) 402 pp. 751–771.

21. K. J. Lohmann, A. O. Willows, and R. B. Pinter (1991) 'An identifiable molluscan neuron responds to changes in earth-strength magnetic fields' *J. Exp. Biol.* 161 pp. 1–24; V. Krauthamer (1990) 'Modulation of conduction at points of axonal bifurcation by applied electric fields'. IEEE *Trans. Biomed. Eng.* 37 pp. 515–519.

22. H. Lai, M. A. Carino, and I. Ushijima (1998) 'Acute exposure to a 60 Hz magnetic field affects rats' water-maze performance'. *Bioelectromagnetics* 19 pp. 117–122.

23. C. K. Chou, K. C. Yee, and A. W. Guy (1985) 'Auditory response in rats exposed to 2,450 MHz electromagnetic fields in a circularly polarized waveguide' *Bioelectromagnetics* 6 pp. 323–326.

24. D. Concar (1999) 'Special Investigation: Mobile Phones' *New Scientist* 10 April 1999 pp. 20–23; D. M. Hermann and K. A. Hossmann (1997) 'Neurological effects of microwave exposure related to mobile communication' *J. Neurol. Sci.* 152 pp. 1–14; R. J. Reiter (1993) 'Static and extremely low frequency electromagnetic field exposure: reported effects on the circadian production of melatonin' *J. Cell Biochem.* 51 pp. 394–403; R. Sandyk (1994) 'Improvement in word-fluency performance in Parkinson's disease by administration of electromagnetic fields' *Int. J. Neurosci.* 77 pp. 23–46; R. Sandyk (1994) 'Alzheimer's disease: improvement of visual memory and visuoconstructive performance by treatment with picotesla range magnetic fields' *Int. J. Neurosci* 76 pp. 185–225; R. Sandyk (1997) 'Resolution of sleep paralysis by weak electromagnetic fields in a patient with multiple sclerosis' *Int. J. Neurosci.* 90 pp. 145–157; W. H. Bailey, S. H. Su, T. D. Bracken, and R. Kavet (1997) 'Summary and evaluation of guidelines for occupational exposure to power frequency electric and magnetic fields' *Health Phys.* 73 pp. 433–453; C. Eulitz, P. Ullsperger, G. Freude, and T. Elbert (1998) 'Mobile phones modulate response patterns of human brain activity' *Neuroreport* 9 pp. 3229–3232.

25. P. A. Valberg, R. Kavet, and C. N. Rafferty (1997) 'Can low-level 50/60 Hz electric and magnetic fields cause biological effects?' *Radiat. Res.* 148 pp. 2–21.

26. J. F. Nielsen (1996) 'Repetitive magnetic stimulation of cerebral cortex in normal subjects' *J Clin Neurophysiol* 13 pp. 69–76; R. R. Ji, T. E. Schlaepfer, C. D. Aizenman, C. M. Epstein, D. Qiu, J. C. Huang, and F. Rupp (1998) 'Repetitive transcranial magnetic stimulation activates specific regions in rat brain' *Proc. Natl. Acad. Sci. USA* 95 pp. 15635–15640; R. Q. Cracco, J. B. Cracco, P. J. Maccabee, and V. E. Amassian (1999) 'Cerebral function revealed by transcranial magnetic stimulation' *J. Neurosci Methods* 86 pp. 209–219; R. M. Post, T. A. Kimbrell, U. D. McCann, R. T. Dunn, E. A. and S. R. Weiss (1999) 'Repetitive transcranial magnetic stimulation as a neuropsychiatric tool: present status and future potential' *JECT* 15 pp. 39–59; M. S. George, S. H. Lisanby, and H. A. Sackeim (1999) 'Transcranial magnetic stimulation: applications in neuropsychiatry' *Arch. Gen. Psychiatry* 56 pp. 300–311.

27. Richard Feynman (1985) *QED: The Strange Theory of Light and Matter* p. 24.

28. Roger Penrose (1995) *Shadows of the Mind* p. 337.

29. John Gribbin (1984) *In Search of Schrödinger's Cat* p. 171.

30. Voltage-gated channels in neurones have a loop of protein that acts like a kind of automatic plug that swings into place to close the channel after it has it has been open for about one millisecond.

BIBLIOGRAPHY

CHAPTER 1

Aristotle. *Progression of Animals*, trans. E. S. Forster (Harvard Heinemann: London and Cambridge MA, 1968).
Aristotle. *Parts of Animal*, trans. A. L. Peck (Harvard Heinemann: London and Cambridge MA, 1968).
Aristotle. *Metaphysica*, 1049b6.

CHAPTER 2

Attenborough, D. (1984). *The Living Planet* (London: Collins/BBC Books).
Gross, M. (1998). *Life on the Edge: Amazing Creatures Thriving in Extreme Environments.* (New York: Plennum Press).
Madigan, M. T., Martinko, J. M., and Parker, J. (1997). *Brock: Biology of Microorganisms* (Upper Saddle River, New Jersey: Prentice Hall International, Inc.).
Postgate, J. R. (1992). *Microbes and Man* (Cambridge: Cambridge University Press).
Postgate, J. R. (1994). *The Outer Reaches of Life* (Cambridge: Cambridge University Press).
Wilson, E. O. (1994). *The Diversity of Life* (London: Penguin Books).

CHAPTER 3

Alberts, B., Bray, D., Lewis, J., Raff, M., Roberts, K., and Watson, J. D. (1994). *Molecular Biology of the Cell* (New York: Garland Publishing, Inc.).
Magner, L. N. (1979). *A History of the Life Sciences* (New York: M. Dekker).

Schrödinger, E. (1944). *What is Life?* (Cambridge: Cambridge University Press).

Singer, C. (1959). *A History of Biology to About the Year 1900* (London: Abelard-Schuman).

Smith, C. U. M. (1976). *The Problem of Life: An Essay on the Origins of Biological Thought.* (London: Macmillan).

Watson, J. (1968). *The Double Helix* (London: Weidenfeld and Nicolson).

Watson, J. D., Hopkins, N. H., Roberts, J. W., Steitz, J. A., and Weiner, A. M. (1987). *Molecular Biology of the Gene* (Menlo Park: Benjamin/ Cummings).

CHAPTER 4

Barrow, J. D. (1988). *The World Within the World* (Oxford: Oxford University Press).

Barrow, J. D. and Tipler, F. (1999). The *Anthropic Principle* (Oxford: Oxford University Press).

Cairns-Smith, A. G. (1985). *Seven Clues to the Origin of Life* (Cambridge: Cambridge University Press).

Darwin, C. (1859). *The Origin of Species.* (London: Penguin Classics, 1985).

Darwin, C. (1944). *Autobiography of Charles Darwin* (London: Watts & Co.).

Davies, P. (1998). *The Fifth Miracle: The Search for the Origin of Life.* (London: Penguin Books).

Dawkins, R. (1976). *The Selfish Gene* (Oxford: Oxford University Press).

Dawkins, R. (1982). *The Extended Phenotype* (Oxford: Oxford University Press).

Dawkins, R. (1988). *The Blind Watchmaker* (Oxford: Oxford University Press).

Dawkins, R. (1996). *Climbing Mount Improbable* (London: Penguin Books).

Dennet, D. C. (1995). *Darwin's Dangerous Idea* (New York: Simon and Shuster).

Fortey, R. (1997). *Life: An Unauthorised Biography* (London: HarperCollins).

Gould, S. J. (1991). *Bully for Brontosaurus* (New York: W.W Norton and Company).

Gould, S. J. (1993). *Eight Little Piggies* (London: Penguin Books).

Gould, S. J. (1996). *Dinosaur in a Haystack* (London: Jonathan Cape).

Kauffman, S. (1995). *At Home in the Universe* (London: Penguin Books).

Levy, S. (1992). *Artificial Life: The Quest for a New Creation* (London: Penguin Books).

Maynard Smith, J. (1993). *Did Darwin Get It Right?* (London: Penguin Books).

Maynard Smith, J. (1993). *The Theory of Evolution* (Cambridge: Cambridge University Press).

Orgel, L. E. (1973). *The Origins of Life* (London: Chapman and Hall).

Waldrop, M. M. (1992). *Complexity* (London: Penguin Books).

CHAPTER 5

Brady, J. E. and Holum, J. R. (1993). *Chemistry* (New York: John Wiley).

Fersht, A. (1985). *Enzyme Structure and Mechanism* (New York: W. H. Freeman).

Rose, S. (1991). *The Chemistry of Life.* (London: Penguin).

Zubay, G. L. (1996). *Biochemistry* (Dubuque, IA: McGraw-Hill Companies, Inc.).

CHAPTER 6

Atkins, P. W. (1994). *The Second Law: Energy, Chaos and Form* (New York: Scientific American Books).

Carnot, S. (1960). *Reflections on the Motive Power of Fire/by Sadi Carnot/and other Papers.* (New York: Dover Publications).

Feymman, R. P. (1985). *QED: the Strange Theory of Light and Matter* (London: Penguin Books).

Monod, J. (1972). *Chance and Necessity* (London: Collins).

Prigogine, I. (1980). *From Being to Becoming* (San Francisco: W. H. Freeman and Company).

Silver, B. L. (1998). *The Ascent of Science* (Oxford: Oxford University Press).

Wolfson, R. and Pasachoff, J. A. (1995). *Physics with Modern Physics: for Engineers and Scientists* (London: HarperCollins College Publishers).

CHAPTER 7, 8, 9

Albert D. Z. *Quantum Mechanics and Experience* (Harvard: Harvard University Press; 1994).

Davies P. C. W.; Brown J. R. *The Ghost in the Atom* (Cambridge: Cambridge University Press; 1986).

Feynman R. P.; Leighton R. B.; Sands M. *The Feynman Lectures on Physics* (Reading, Massachusetts: Addison-Wesley Publishing Company; 1963).

Feynman R. P. *QED: the Strange Theory of Light and Matter* (London: Penguin Books; 1985).

Gribbin J. *In Search of Schrödinger's Cat* (Reading: Black Swan; 1992).

Gribbin J. *Schrödinger's Kittens and the Search for Reality* (London: Phoenix; 1996).

Lindley D. *Where Does the Weirdness Go?* (London: Vintage; 1997).

Penrose R. *The Emperor's New Mind* (London: Vintage; 1989).

Penrose R. *Shadows of the Mind* (London: Vintage; 1994).

Rae A. *Quantum Physics: Illusion or Reality?* (Cambridge: Cambridge University Press; 1986).

Rae A. *Quantum Mechanics* third ed. (London: The Institute of Physics; 1996).

Wheeler J. A.; Zurek W. H. *Quantum Theory and Measurement* (Princeton: Princeton University Press; 1983).

CHAPTER 13

Bear, M. F., Connors, B. W., and Paradiso, M. A. *Neuroscience: Exploring the Brain* (Baltimore: Williams & Wilkins; 1996).

Cairns-Smith, A. G. *Evolving the Mind: On the Nature of Matter and the Origin of Consciousness* (Cambridge: Cambridge University Press; 1996).

Calvin, W. *How Brains Think* (London: Phoenix; 1996).

Chalmers, D. *The Conscious Mind* (New York: Oxford University Press; 1996).

Dennett, D. *Consciousness Explained* (London: Penguin Books; 1992).

Hardcastle, V. G. *Locating Consciousness* (Philadelphia: J. Benjamins Pub. Co.; 1995).

Libet, B. *Neurophysiology of Consciousness* (Boston: Birkhäuser); 1993.

Nunn, C. *Awareness: What is it, What it Does.* (London: Routledge; 1996).

Penfield, W. *The Mystery of the Mind: A Critical Study of Consciousness and the Human Brain* (Princeton: Princeton University Press; 1975).

Penrose, R. *The Emperor's New Mind* (London: Vintage; 1989).

Penrose, R. *Shadows of the Mind* (London: Vintage; 1994).

Zohar, D. *The Quantum Self* (London: Flamingo; 1991).

INDEX

Page numbers in italics denote an illustration